Patrick Moore's Practical Astronomy Series

Other Titles in This Series

Spectroscopy: The Key to the Stars

Reading the Lines in Stellar Spectra

Keith Robinson B.A., Ph.D., F.R.A.S.

With 78 Figures

Keith Robinson
Royal Astronomical Society, UK
4 Bedford Place
Scotforth, Lancaster, UK

British Library Cataloguing in Publication Data
A catalogue record for this book is available from the British Library

Library of Congress Control Number: 2006930106

Patrick Moore's Practical Astronomy Series ISSN 1617-7185

ISBN-10: 0-387-36786-1   eISBN-10: 0-387-68288-0
ISBN-13: 978-0-387-36786-6   eISBN-13: 978-0-387-68288-4

Printed on acid-free paper
© Springer-Verlag London Limited 2007

9 8 7 6 5 4 3 2 1

Springer Science+Business Media
springer.com

Acknowledgements

My grateful thanks go to Dr. Harry Blom and all at Springer, New York, for their help and enthusiastic support during the writing of the book. Many thanks are also due to John Watson, particularly for his enthusiasm over the original idea for the book. Finally, my heartfelt gratitude (and sympathy) goes to my wife Elizabeth for reading the manuscript and offering many helpful comments and suggestions.

Diagrams

All diagrams were prepared by the author.

Contents

Contents

Introduction

Recently, I attended my usual local astronomy club meeting; our speaker was a local amateur who talked about CCD astronomy. Listening to his talk it seemed to me that here was the very guy for whom I'd written this book. He'd started out by doing CCD photography and had produced images that most of us in the audience could only envy. Later he'd moved onto variable star CCD photometry; it was clear that he wanted to combine his hobby with doing scientifically useful observations. Finally, he said that his most recent adventure was to get into CCD spectroscopy, at this point his talk took on a significantly elevated sense of importance. I asked him about the kind of spectral resolution which amateurs could achieve and he replied enthusiastically that very soon, sub-angstrom resolution would be fairly normal. I admit I was somewhat relieved, because it meant that much of what I've written in this book would not be merely of academic interest.

My own interest in astronomy started in the 1960s; I was just a kid and couldn't get enough of the stuff, but there were two areas which I seemed to avoid like the plague. One of these areas was radio astronomy; in those days it seemed to be more to do with electronics than stars and of course we didn't have the wonderful radio images of today, just wiggly lines on a piece of chart recorder paper. The other area to be avoided was spectroscopy; the physics involved seemed just too advanced. I knew that spectral lines told us about different chemical elements in stars and that red- or blue-shifted lines were caused by the motion of astronomical bodies but that was about it. In any case, spectroscopy was definitely not on the observing agenda of most amateur astronomers.

Technology has changed all that; now amateur astronomers can use CCD cameras together with backyard spectroscopes to do astronomical spectroscopy. Unfortunately, the physics hasn't changed and to be honest, much of it isn't the kind of physics you're likely to do in high school. Virtually all books on astronomical spectroscopy are textbooks which as they say 'don't take prisoners'; they're the kind of books which wouldn't even be used by most undergraduate astronomy students. As for books on spectroscopy aimed at amateur astronomers, aside from Stephen Tonkin's *Practical Amateur Spectroscopy* also published by Springer, they're pretty thin on the ground. A.D. Thackeray wrote a semi-popular book *Astronomical Spectroscopy* back in 1961 but this is long out of print.

Tonkin's book is the first of I hope many in this new area of amateur astronomical research and as its title suggests, it is a practical book. There clearly is a need for a theory book on astronomical spectroscopy which doesn't expect the reader to be a science graduate and I hope this book will at least make a start in fulfilling that need. This is a book about physical processes and physics processes which make astronomical spectra they way they are, so there's quite a bit of physics in it. The only assumption I've made here is that you did some physics in high school; even so, everything is done from the bottom up, so to speak. I've also used a bit of mathematics but only in the form of simple 'plug the numbers in' equations which can easily be done with a pocket calculator. These equations involve no more than knowing about powers of 10; and if this presents a problem, have a read through Appendix A, which should get you up to firing speed. Besides, giving you a feel for what's going on, these simple equations can be used to explore some areas of physics or spectroscopy; that's why they are here. Anything else that can be done without the math **is** done without the math and that means most things.

There are one or two topics which may be unfamiliar and at first may seem a bit involved, but take your time; read things through carefully, more than once if necessary. My real hope is that this book will help you to understand more about what's going on in your spectra and most of all that you'll quickly realise that there's far more to astronomical spectroscopy than just identifying lines in a spectrum. It truly is fascinating stuff and I reckon, I can be pretty certain that you'll want to know more.

Keith Robinson
Lancaster, UK

Spectroscopy—A New Golden Age for Amateur Astronomy

It was the year 1824. Gustav Kirchoff was born. He, together with Robert Bunsen was to lay the foundations of modern spectroscopy. By perhaps an 'inevitable' coincidence, it was in the following year that the French philosopher August Comte pronounced that the chemical composition of the stars was knowledge that mankind would never possess. This is the kind of (well known) story that is guaranteed to make every scientist—astronomer or otherwise—smile. By the 1860s Father Angelo Secchi was classifying stars according to their spectra and astronomical spectroscopy was born.

The latter part of the nineteenth century was also a golden era for amateur astronomy; though it has to be admitted that in those days amateur astronomers usually had money and plenty of it. They were though, blessed with equipment—telescopes and telescope 'add-ons' like bifilar micrometers which were very much on a par with the kind of stuff used by the professionals of the time. The observatory inventory of a typical 'gentleman amateur astronomer' would very likely also include a direct vision spectroscope; this was an arrangement of small prisms and lenses fitted into a tube, which could be inserted into the eyepiece end of the telescope. This enabled the spectra of interesting stars to be examined and 'commented on'; the fact is that at this time not a great deal was understood about the mysterious lines in the spectra of stars, except of course that some of them could be identified with certain chemical elements. Sir William Huggins had even realised that small shifts observed in the lines when compared to laboratory spectra, resulted from a star's motion towards or away from the Earth; but an explanation for exactly how the lines were formed and why they appeared to form the patterns they did, had to wait until the early decades of the twentieth century.

As the twentieth century itself progressed, that golden age of amateur astronomy was gone too; professional astronomy was advancing at a furious pace both in terms of equipment and theory. The amateurs, even the wealthy ones couldn't hope to match in size telescopes like the 100 inch on Mt. Wilson (even if they had the money it wouldn't have fit into their backyards); but what was perhaps worse for the amateur was that much of professional astronomy had evolved into astrophysics and this was very much to do with the rising importance of spectroscopy in astronomy. Tremendous progress in theoretical physics; most importantly quantum mechanics, had honed observational spectroscopy into the most formidable analytical tool for professional astronomers; but quantum mechanics itself in those days was causing problems even for the big names in physics including Albert Einstein. It was weird stuff which did not sit easy, if at all with 'common sense' everyday thinking about the world and if this was how it was for professional scientists, then the amateur astronomers were bound to get left behind.

However, the role of amateur astronomers didn't get lost altogether thanks in no small measure to the efforts of a few dedicated people and also to the fact that the Universe is after all a pretty big place; big enough for the 'small kids on the block' to do their bit too. This became perhaps more apparent than anywhere else with the discovery of more and more variable stars. What started out as a biggish handful towards the end of the nineteenth century had grown to thousands by the mid-twentieth century. There were just too many of them for the professional astronomers to deal with so here the dedicated amateur was not only welcomed but positively needed by the professionals. I reckon one of the best things that a seasoned astronomer, professional or amateur can tell a beginner about, is the superb work done by amateur variable star observers all over the world. Their observations don't get filed away in scrapbooks; they are used very gratefully by professionals and get published in research papers all the time. The amateur's role here is usually that of an unsung hero but the knowledge that you're contributing to astronomical science is immensely fulfilling; it's great to feel needed.

Now however time has moved on again and technology has moved even faster but it has become cheaper—very much cheaper. Once again the amateur who this time doesn't have to be as wealthy in real terms as his nineteenth century counterpart can use equipment and indeed, thanks to the Internet, resources like databases, which compare with those used by the professionals. He/she can't compete in terms of telescope size but the modern day dedicated amateur can develop what amounts to a professional class observatory in miniature, and as I mentioned above, the Universe is so big that there's plenty of scope for anyone who wants to make themselves useful to the astronomical community. This now includes doing spectroscopy; until recently a more or less forgotten skill among amateurs but now growing again thanks to CCD's and 'off the peg' affordable spectroscopes. It's probably a safe bet to say that in the very near future, organisations like the American Association of Variable Star Observers (AAVSO) will develop spectroscopy programs to tie into their variable star photometry programs. Professional astronomers use photometry and spectroscopy side by side so it surely follows that if amateurs do the same the value of their work will truly be enormous.

There is perhaps one big snag though; the theory behind those spectral lines hasn't gone away; it involves all that weird quantum mechanics not to mention the mathe-matics. This is probably a big turn off to many potential amateur spectroscopists. If you were to ask amateur astronomers about which areas of astronomy they know and understand least, spectroscopy would be sure to figure high up the table. Try reading a professional astronomy research paper involving spectroscopy; it will certainly include a lot of jargon and terminology; involve some pretty high-level (at least undergraduate level) physics; and then there's the maths! Virtually all the books on astronomical spectroscopy are written at the undergraduate or postgraduate level; so as far as the average amateur is concerned these books 'don't take prisoners'. If it were possible to understand spectroscopy better without the need for a physics degree, I reckon more would be drawn to this pioneering area of amateur astronomy. Even the experienced amateur would surely benefit from having a better understanding as to why, for example; some spectral lines were deeper or broader than others; why some were double peaked; and why do the spectra of molecules look so complicated?

This is where I hope this book will at least make a start; to give you some understanding of the 'whys' and 'hows' of astronomical spectra. This book is about the physical processes and the physics processes going on in stars and nebulae, which make their spectra the way they are; it's not a 'train spotter's guide' to the lines in stellar spectra. This means that we'll be learning some physics—notably the weird stuff; but it really is possible to gain a sound insight into quantum mechanics without the mathematics and without a physics degree. As spectroscopists we only need to know how quantum mechanics works within atoms and provided we're prepared to accept a few of the most basic ideas about quantum theory (even the professionals have to do this), everything else will be seen to follow with breathtaking logic.

We won't be needing any 'heavy mathematics' either, however, there are some very simple 'plug the numbers in' equations, or formulae which can be used very easily to work out very useful numbers. A good example is a simple formula which enables us to convert the temperature of a gas into the speed of its atoms and this can then be used to tell us how broad a spectral line is likely to be. Simple things like this, which can be easily done with a pocket calculator can make amateur spectroscopy much more interesting and rewarding. Simple calculations like this will of course be explained very fully with step-by-step instructions on how to make sure you get the right answers.

Once we've covered the basic material the rest of the book will introduce some of the most important phenomena which are seen in astronomical spectra and explain how they work. This will involve using astronomical objects like gaseous nebulae or accretion disks as 'spectral laboratories' where we can see how the physics works. However, the spectra themselves are the real 'stars' rather than the astronomical objects; so for example, when we use cool red stars to introduce molecular spectra, molecular spectra is what that chapter is really all about rather than cool red stars.

Hopefully by the time you reach the end, you'll be able to understand more about what's going on in your spectra; they are after all potentially priceless data in this new golden age of amateur astronomy.

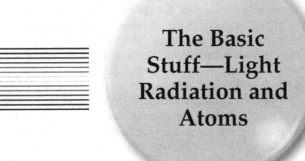

The Basic Stuff—Light Radiation and Atoms

A modern astronomical spectrum is basically a graph that plots intensity of radiation, for example visible light, against wavelength or frequency. The fascinating character of astronomical spectra is the result of light and other forms of electromagnetic radiation interacting with matter—i.e. atoms and molecules. In this chapter, we'll review some basic things about light and electromagnetic radiation. We'll take a look at the purest spectrum; the so-called 'black-body' spectrum and how nineteenth-century physics failed to explain it. The consequences of its solution, a 'quick fix' at the time turned out to be profound to say the least and resulted in the whole science of spectroscopy as we know it today. Then we'll look at the structure of atoms and see how there were problems here too, which could only be solved by introducing what at the time were some pretty weird ideas.

Light

Light has a dual personality and we need to be aware of and indeed comfortable with both aspects of its character when we learn about spectra. The debate about the nature of light was really hotting up in the nineteenth century when experiments seemed to show that light was some form of wave motion rather like waves or ripples on water. Isaac Newton had previously favoured the idea that light consisted of a stream of energetic particles or corpuscles.

Light Waves

In 1864, the Scottish theoretical physicist James Clerk Maxwell described the precise mathematical relationship between electric fields and magnetic fields. These fairly abstract concepts enable forces to act across empty space just as gravity does; Albert Einstein even tried to explain them as some property of curved space-time just as he had done so brilliantly with gravity itself. He failed however and to this day no one can really say what an electric or a magnetic field actually is; physicists can only describe how they work. They are though, familiar from everyday life; the Earth's magnetic field makes our compasses work and electric fields produce bolts of lightning. The source of a simple electric field is *electric charge*, which is a basic property of matter; thus in the space surrounding every electric charge, there is an electric field. Electric charge comes in two forms, 'positive' and 'negative' and as they often say about people, 'likes' repel and 'unlikes' attract. Positive charges repel each other and so do negative charges. In turn, positive charges attract negative ones. The force of attraction or repulsion between electric charges is truly enormous but because of the dual nature of electric charge, positive and negative electric charges effectively cancel each other out. Most things on planet Earth including you and I are an even mixture of positive and negative charge so we are effectively electrically neutral. The final thing to say about electric fields here is that they have a direction in space; the field points in the direction in which it would make a positive charge move.

A magnetic field also has a direction but there is no 'magnetic charge'; instead there is the concept of a *magnetic pole*. These always come in pairs; a north pole and a south pole which are best thought of as being at the opposite ends of a bar magnet. Again, rather like electric charges, like poles repel and unlike poles attract. The direction of

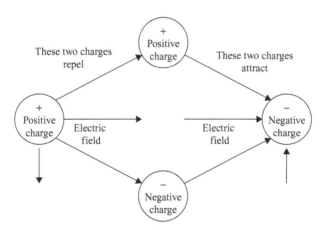

Figure 2.1. Electric charge is a fundamental property of matter; two positive or two negative charges will repel each other across empty space whereas a positive and a negative charge will attract each other. The electric field surrounding an electric charge has a direction, i.e. the direction in which a positive charge will move under the action of the field.

a magnetic field is the direction in which a fictitious isolated north pole would move under its influence.

Maxwell showed that for electric and magnetic fields, as the song goes, 'you can't have one without the other'. If an electric field changes; for example, as a result of electric charge moving, a magnetic field is produced; electric charge moving along a wire constitutes an electric current, which gives rise to a varying electric field, and this in turn creates a magnetic field surrounding the wire. A changing magnetic field produces an electric field; a dynamo generates electricity as a result of a coil of wire being made to turn in a surrounding magnetic field. This creates an electric field, which drives an electric current along the wire. This dynamic interplay between electric and magnetic fields turned out to be truly magical and produced the proverbial 'rabbit' that came out of Maxwell's 'hat'. Maxwell discovered that varying electric and magnetic fields could move together as waves through space. This happens when electric charges, the source of an electric field, accelerate, i.e. *change their velocity* and one very important example of an accelerating charge is one which vibrates or oscillates. Oscillating electric charges then produce oscillating electric and magnetic fields which in turn move outwards as waves. Maxwell was able to derive a very simple formula for the speed of these *electromagnetic waves* and this speed came out to be equal to the speed of light, which itself had been determined by this time. So Maxwell concluded that light itself consists of electric and magnetic fields propagating together as waves, i.e. electromagnetic waves. Maxwell thus provided the theory that vindicated what the experiments of the time seemed to show—light was made of electromagnetic waves.

We've all seen waves on the surface of water. A water wave consists of a succession of ripples or crests and troughs that appear to spread out across the surface of the water. One of the most important quantities associated with a wave like this is the distance between two neighbouring crests or troughs. For a steady wave this distance doesn't change; it is called the *wavelength* and is always denoted by the Greek letter 'lambda' or 'λ'.

A light wave is not so easy to visualise; we can't actually 'see' electric and magnetic fields but their strengths and directions can be measured. If we were to measure the strength and direction of say, the electric field at some point in a beam of light, we would find first of all that its direction was the same as the direction along which the

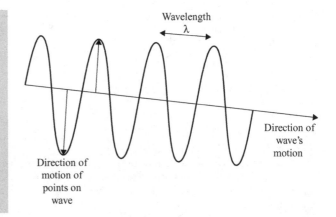

Figure 2.2. A simple transverse wave such as a water wave. The motion of every point on the wave is always at right angles to the direction in which the wave is travelling. For a steady wave, the wavelength is constant.

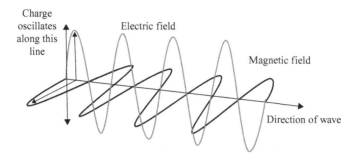

Figure 2.3. A crude representation of an electromagnetic wave; the plane of the 'electric field wave' coincides with the line along which the charge oscillates. The plane of the 'magnetic wave' is at right angles to this and the direction in which the wave travels is at right angles to both the electric and the magnetic fields. An electromagnetic wave is thus emitted at right angles to the line of charge oscillation and never along this line.

charge was oscillating and was always at right angles to the direction in which the light wave was travelling. Physicists call this kind of wave a *transverse wave*. The electric field at any point on the wave would first point in one direction, rise to a maximum value then decline to zero. It would then increase, pointing in the opposite direction, rise to a maximum again and finally decline back to zero. The associated magnetic field would be doing the same thing and we would also find that the direction of the magnetic field was at right angles to both the electric field and the direction in which the wave was travelling.

A final point to note is that the electromagnetic wave 'comes out' at right angles to the line along which the charge oscillates and never along the direction of this line; this point will be significant later on in Chapter 11.

For a light wave, these processes happen very rapidly; the number of times the electric or magnetic field oscillates in 1 s is another very important number associated with a wave—*the frequency*, denoted by the Greek letter 'nu' or 'ν'. The wavelength and frequency of a light wave are connected by a very simple formula:

$$c = \nu \times \lambda. \tag{2.1}$$

Here c is the speed of light, i.e. the speed at which the electromagnetic wave travels through space. The speed of light is constant so what this formula says is simply that the shorter the wavelength, the higher the frequency and vice versa.

Another way of thinking about this is that if we make an electric charge oscillate, electromagnetic waves will be produced. If the charge oscillates relatively slowly, i.e. with a low frequency, then the frequency of the resulting waves will be low and the wavelength long. Speed up the frequency of the oscillating charge and the frequency of the electromagnetic waves increases together with a corresponding decrease in wavelength. It takes more energy to make an electric charge oscillate more rapidly so we'd expect the resulting electromagnetic waves to have more energy. In other words, we expect waves with a shorter wavelength or higher frequency to have more energy than those with a lower frequency.

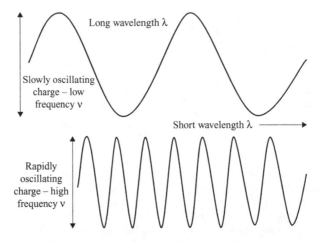

Figure 2.4. A slowly oscillating electric charge generates a wave of long wavelength; a rapidly oscillating charge (higher frequency) generates a short wavelength wave. The same kind of thing happens when you make waves on a length of rope, one end of which is tied to a tree.

What kind of numbers are we in fact talking about here? The speed of light is known to be very nearly equal to 300,000 km/s or 3×10^5 km/s. (In case you need a quick refresher on powers of 10, see Appendix A.) How you talk about wavelength may depend on whether you're a physicist or an astronomer; physicists always like to measure lengths in metres. Visible light has a very short wavelength indeed; of the order of a few hundred *nanometers*. A nanometre is one thousand millionth of a metre or 10^{-9} m. So, a physicist might speak of light having a wavelength of say 500 nm. When speaking about visible light, astronomers use a different unit of distance called the *angstrom* (named after Anders Jonas Angstrom who ironically was a physicist from Sweden), which is represented by the symbol 'Å'. One angstrom equals 10^{-10} m, so 10 Å make a nanometre. Thus, the wavelength of visible light is of the order of several thousand angstroms. In a typical astronomical journal or research paper, a wavelength would be written as λ4686 or λλ4000 to 5000 meaning 4686 Å (468.6 nm) and 'in the range' 4000 to 5000 Å (400 to 500 nm), respectively. The human eye is most sensitive to light with a wavelength of about 5000 Å.

From Eq. (2.1), we can easily calculate the frequency of a 5000 Å beam of light; in other words, how rapidly the electric and magnetic fields are oscillating. First we need to do things the physicist's way and convert all distances to metres; 5000 Å equals 5000×10^{-10} m which equals 5×10^{-7} m. The speed of light converts from 3×10^5 km/s to 3×10^8 m/s. The frequency is found by dividing the speed of light by the wavelength, i.e. 3×10^8 divided by 5×10^{-7}. This gives 6×10^{14} which is the number of times the electric and magnetic fields are oscillating each second. This is an enormous number and in fact when talking about different parts of a spectrum, it's much more common to use wavelength, if for no other reason than wavelengths which involve numbers like 5003 or 6563 become much more familiar than those like 6×10^{14}. Frequency does come up sometimes in the literature though and it is used quite a lot by physicists in particular, mainly because the energy associated with a light wave is directly related to the frequency in a particularly simple way as we shall see later. So you need to be aware of this, as well as how to convert frequencies to wavelengths. Another quantity that is sometimes used is the reciprocal or one divided by the wavelength; this is called the *wave number* because it gives the number of complete waves in a standard distance. For this the wavelength has to be not in angstroms or nanometres but in centimetres or metres; for example, 5000 Å is 5×10^{-5} cm and one divided by this equals 2×10^4 or 20,000 waves/cm (try this on your pocket calculator). So the units of wave number are 'per cm' (cm^{-1}) or 'per metre' (m^{-1}).

Light and Colour

Isaac Newton carried out experiments with prisms that showed that white light is a combination of all the colours of the visible spectrum. Prisms in fact work because the speed of light in glass is different for different colours; blue light travels more slowly in glass than red light. The result is that a beam of blue light striking the face of a prism at an oblique angle will have its path deflected as it enters the glass; what's more, this deflection is greater than would be the case for a beam of green light. This in turn would be greater than that for yellow light and so on. A prism thus separates out the various colours from a beam of white light; this process is called *dispersion*.

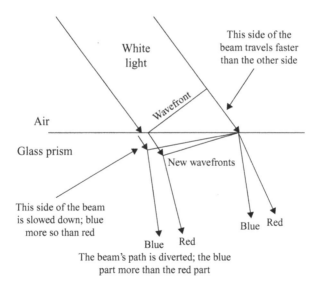

Figure 2.5. When a beam of white light passes from air into a denser optical medium (e.g. a glass prism) it slows down; blue light slows more than red light. If the beam hits the glass surface at an angle other than 90°, the blue part of the beam is diverted more than the red part (other colours fall between these two extremes); the white light is dispersed and forms a spectrum. The wavefronts are always at right angles to the light beam's direction.

Once the wave nature of light was established, it was realised that the colour of a beam of light is determined by its wavelength. Blue light has a shorter wavelength than red light, and green and yellow light, etc. have wavelengths somewhere in between. Because the wavelength can vary continuously from one end of the spectrum to the other, there are clearly an infinite number of colours in the visible spectrum and not just the seven identified by Newton. The visible spectrum runs from about λ4000 at the violet end to about λ7000 at the red end. As mentioned above, the human eye is most sensitive to light at about λ5000 and this lies in the green part of the spectrum.

Electromagnetic Radiation

Electromagnetic waves can have wavelengths less than λ4000 and greater than λ7000. These other wavelength regions beyond the violet and red ends of the visible spectrum make up the whole *electromagnetic spectrum or 'e-m' spectrum.* Fig. 2.6 shows the various regions of the electromagnetic spectrum and there are a couple of points to note; the visible part of the spectrum is really a pretty narrow region of the whole thing. There is also of course no sharp dividing boundary between the different regions; ultraviolet radiation merges into X-rays which in turn merge into gamma rays and so on. If you're an amateur astronomer who does spectroscopy, you'll almost certainly be

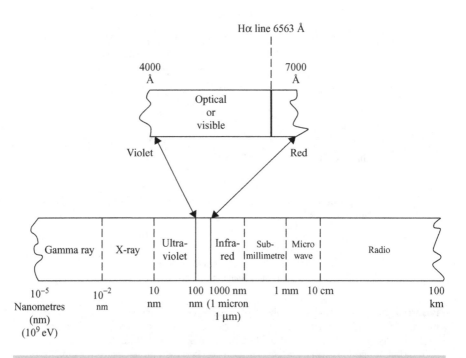

Figure 2.6. The electromagnetic spectrum; the numerical scale at the bottom is not 'done to scale'.

concerned exclusively with the visible part of the spectrum. However, it's important to realise that professional astronomers now do spectroscopy in just about every region of the electromagnetic spectrum. So if you're browsing the literature, you're certain to come across spectra taken in maybe the ultraviolet or infrared regions. Such spectra look just like visible light spectra and indeed many processes and phenomena, which affect visible light spectra, also apply to other parts of the e-m spectrum. Their acquisition however requires methods and equipment (telescopes in space for example), which at the present time lie for the most part (just?) beyond the amateur's grasp. We will talk about spectra in the non-visible region of the e-m spectrum when it is relevant.

One thing you're certain to come across when reading articles or papers in journals is the way that astronomers talk about wavelength when discussing different parts of the e-m spectrum. Angstroms are still used for ultraviolet wavelengths and to some extent longer wavelength X-rays; for short wavelength or 'harder' X-rays and gamma rays you're more likely to come across the term *'electronvolt or eV'* which is not a wavelength at all but a measure of energy. As we move into the infrared part of the electromagnetic spectrum, angstroms give way to microns (short for micrometres), a micron being one millionth of a metre. As wavelengths increase we enter the *sub-millimetre* region and if we happen to be a radio astronomer we talk about wavelengths in centimetres and metres.

The Beginning of Spectroscopy

Newton's experiments with narrow beams of sunlight and prisms were carried out in a darkened room and showed the familiar rainbow of colours which make up white light. Newton saw no more detail in the solar spectrum than this which may in part have been due to the quality of glass in the prisms he was using. In 1814 however, Joseph von Fraunhofer using his own self-made prisms observed several hundred dark lines crossing the Sun's spectrum. He determined the wavelength of some of the lines and labelled the most prominent ones from A to K. This notation is still used today in relation to the solar spectrum and of course Fraunhofer has the honour of having the lines in the spectrum of the Sun named after him. What of course he did not know was what caused them.

One area of science that was in its heyday in the nineteenth century was chemistry. In fact, it was the chemists rather than the physicists who were gathering evidence that all matter was made of tiny particles called *atoms*. Their experiments showed that certain substances could only react with a definite and fixed amount of some other substance and the only way that this could be satisfactorily explained was by assuming that the stuff in their test tubes was made of atoms which were bonded together to make *molecules*. One of the great achievements of nineteenth-century chemistry was the building up of *the periodic table of the elements*. Elements are substances, which cannot be broken down into anything simpler by chemical processes; common examples are iron and sulphur. Atoms of elements can chemically 'bond' together to make molecules, which are the basic units of chemical *compounds*. By careful analysis, chemists determined the correct order for elements to be placed in the periodic table; for example, hydrogen was observed to react with other substances in smaller amounts than that for any other element and so it was concluded that hydrogen was the lightest of all the elements and occupied position number one in the periodic table; next came helium, lithium, beryllium, boron, carbon, etc. As we shall see, the physicists in turn discovered the basic reason why the elements are ordered in the way they are. Whats perhaps the most remarkable thing to come out of all of this is that there are only 92 naturally occurring elements starting with hydrogen and finishing with uranium at position number 92.

If you enjoyed chemistry at high school (I did) you'll remember that one of the most fascinating things was to 'burn' some chemical salt, for example, copper sulphate in the flame of a Bunsen burner and note how the flame acquired often vivid hues of blue-green or red. Still more interesting was to use a spectroscope to look at the spectrum of light given off by vaporised chemical salts. The foundations of modern spectroscopy were laid by Robert Bunsen (of Bunsen burner fame) and Gustav Kirchoff who carried out experiments just like this in the mid-nineteenth century. They discovered that chemical salts invariably produced spectra which consisted of a series of bright lines (these were in fact images of the slit in their spectroscope; each one corresponding to a different wavelength) set against a dark background. This was very much in contrast to the continuous spectrum produced by sunlight. Whats more they realised that each chemical produced its own characteristic set of lines and in fact identifying a set of lines was a way of identifying a particular chemical element. This of course would later prove to be the key to identifying different chemical elements in the stars.

Besides the chemistry aspect of Bunsen and Kirchoffs work, Kirchoff himself produced three laws, now known as *Kirchoff's laws* which were very much to do with the physics of spectra. These laws stated that spectra came in three and only three types, depending on the conditions in which they were produced:

- Continuous—basically a rainbow of colours. This kind of spectrum is produced by a hot incandescent solid or a dense gas. A hot liquid such as molten iron would also produce a continuous spectrum.
- Emission—isolated bright lines at different wavelengths seen against an otherwise dark background. This kind of spectrum is produced by a hot incandescent *thin or low-density* gas. The pattern of these *emission lines* is determined by the chemical nature of the gas producing it.
- Absorption—a continuous spectrum with superimposed dark lines. This spectrum, the sort produced by most stars results from relatively cool thin gas situated between the source of the continuous spectrum and the observer. The pattern of dark lines or *absorption lines* is again determined by the chemical nature of the intervening cool gas.

Bunsen and Kirchoffs work thus provided the key to unlocking the chemical secrets of the stars. In the latter part of the nineteenth century, astronomers were attaching spectroscopes to telescopes; they examined, first visually and eventually photographically the spectra of stars. Most, though not all were seen to be absorption spectra and by comparing the pattern of lines seen in stellar spectra with comparison spectra produced in a laboratory, it was possible to determine those chemical elements which were present in the outer layers of stars. It's well known of course that helium was discovered first in the spectrum of the Sun and most people associate astronomical spectroscopy with the chemistry of the stars but we shall see that

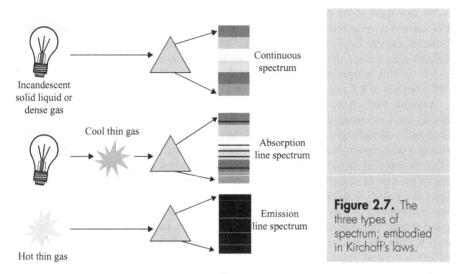

Figure 2.7. The three types of spectrum; embodied in Kirchoff's laws.

as spectroscopy developed in the twentieth century, it became a formidable tool for investigating all manner of dynamic processes which take place in stars and stellar systems.

Black-Body Radiation—The Perfect Spectrum

The work of Bunsen and Kirchoff showed that there are basically three types of spectra; continuous, absorption and emission. A continuous spectrum shows radiation emitted over a whole range of wavelengths; an absorption spectrum is similar but there are gaps where radiation is missing and an emission spectrum consists mainly of gaps with radiation visible at only certain discrete wavelengths. Clearly, the missing bits in an absorption spectrum are caused by the chemical nature of whatever lies between the original source of continuous emission and the observer and in turn the pattern of lines in an emission spectrum is determined by the nature of the hot thin gas producing it. What about the continuous spectrum though? For example is the continuous spectrum produced by a piece of hot iron the same as that produced by a piece of hot titanium? How would we compare one continuous spectrum with another? After all, there are no absorption lines to act as identifying finger prints. The answer lies in the very nature of what a spectrum is; a spectrum is a graph, which shows how the radiation emitted by something is distributed among the various wavelengths of the electromagnetic spectrum. So what would distinguish one continuous spectrum from another would be the shape of the spectrum graph itself. This is indeed found to be the case; different substances when heated produce continuous spectra. The intensity of emitted radiation over different wavelength ranges differs according to the substance under investigation and what's more, for a given substance, the shape of the continuous spectrum can vary as the temperature rises. However, could there be some kind of benchmark spectrum; a sort of idealised distribution of emitted radiation across the various wavelengths?

We normally think of temperature as something which is measured in degrees Celsius or °C; some of us even still use degrees Fahrenheit. Astronomers and physicists however use the *Kelvin* temperature scale; this scale runs in just the same way as the Celsius scale except that zero degrees Kelvin (written as 0K *not* 0°K) lies at about −273°C. This temperature is called *absolute zero*. So 0°C is equal to about 273 K; 100°C equals 373 K and so on.

Any object which is at any temperature above absolute zero is 'warm' and all warm bodies emit electromagnetic radiation. Human beings are warm to the touch; we emit infrared radiation. Stars are much hotter; they emit visible light. The central star of a planetary nebula is very hot indeed and emits a lot of ultraviolet radiation. Objects also absorb radiation; our skin warms up when exposed to sunlight; gas in the Orion nebula absorbs ultraviolet radiation from nearby hot stars and emits visible light as a result. How efficient are objects at absorbing and emitting radiation? This was investigated by Kirchoff and he came to the conclusion that an object which was good at absorbing radiation was also good at emitting radiation. In fact, he went one step further and said that the absorbing efficiency of something (he called this its *absorptivity*) and its emitting efficiency (its *emissivity*) were equal. He postulated the idea of a body, which could with total efficiency; absorb all electromagnetic radiation falling on it. Such a body would then with equal efficiency emit the maximum possible amount of electromagnetic radiation at all wavelengths. Such a body he called a *black body* and radiation emitted by such a body was called *black-body radiation*.

This is quite an abstract idea; physics textbooks usually describe a hollow cavity which is heated from the outside and maintained at a constant temperature so that the inside walls of the cavity emit electromagnetic radiation. The cavity has a tiny hole in it so that any radiation from outside which enters the hole will be fully absorbed by the hole. In turn, radiation from inside the cavity, which escapes via the hole will do so in such a way as to mimic very accurately a black body. So radiation from the hole closely resembles that from a perfect black body.

This is still somewhat abstract; i.e. we have a 'hole' to represent a perfectly radiating body. Sometimes you can get a better feel for an idealised phenomenon by looking at real situations, which are obviously not ideal, because of some complication. If you understand what the result of the complication is, it's easier to see what the result would be if the complication were removed. Take for example the star Vega; just suppose that Vega were in fact a perfect black-body radiator. In the early 1980s, the *Infrared Astronomical Satellite* discovered that Vega was surrounded by a cloud of dust. The evidence for this was what astronomers called an 'infrared excess' in Vega's spectrum. Some of the visible light from Vega is absorbed by the cool surrounding dust; the result is that the dust warms up and re-radiates the absorbed light in recycled form, namely as infrared radiation. So this complication—the dust, has removed part of Vega's hypothetical perfect black-body spectrum, recycled it and superimposed it on another part of the spectrum. This leaves Vega with a non-black-body spectrum.

Of course even without the surrounding dust, Vega does not emit a pure black-body spectrum but it illustrates the point that it is processes like this which cause objects like stars to have spectra which deviate from the ideal. Deep inside a star, matter at least on a local scale does radiate like a black body but there are processes going on in the outer layers of stars, which degrade the spectrum from that of the ideal black body. Having said all this though, it turns out that at least to a first approximation, the continuous part of the spectra of many stars is not too different from that of a black body and in fact over restricted wavelength regions a star may actually radiate as a perfect black body. The perfect black-body spectrum does exist in one place of course—the cosmic microwave background radiation.

A black-body spectrum does not depend on the chemical nature of whatever is producing it; it depends only on the object's temperature. Late nineteenth-century physicists performed experiments, which simulated as closely as possible the radiation from ideal black bodies. They examined and recorded the spectra from such bodies for a range of temperatures. Three things became clear:

- The hotter a body was, the more radiation it gave off. So if plotted on the same scale, the black-body spectrum of a hotter body would be higher up the scale than that from a cooler body.
- The curves were roughly 'bell shaped' having low emission values for long wavelengths, rising to a maximum or peak value and then declining again towards short wavelengths.
- The hotter the body was, the shorter the wavelength at which peak emission occurred.

Fig. 2.8 shows a series of ideal black-body spectra for a range of temperatures. The first and third points make sense because if we heat a body to a higher temperature, we're pumping more energy into it, so we'd expect more to be re-radiated. This simply means that the radiated energy values are higher which results in a spectrum plot which is higher up the intensity scale. Also, it's well known that when an object is heated it will eventually glow red and then orange and finally white with perhaps a blue tinge. So as the body gets hotter, more and more shorter wavelength radiation is being emitted. The bell shape of the black body curve itself was of course experimentally determined but it became an issue of great importance at the end of the nineteenth century to work out theoretically how this curve was produced.

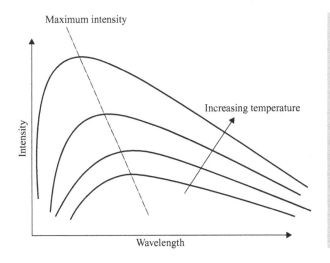

Figure 2.8.
Stylised black body radiation curves for a range of temperatures. The wavelength at which maximum emission occurs shifts towards shorter wavelengths as the temperature increases.

Quantum Theory Is Born

Explaining the ideal continuous spectrum—the black-body spectrum, proved to be one of the biggest problems for physicists at the end of the nineteenth century. The way they saw things was like this; the electromagnetic radiation emitted by a black body is produced by oscillating electric charges near the body's surface. By this time it was generally agreed that electric charge was a fundamental part of the structure of atoms and that the result of heating an object was to make the electric charges in its atoms oscillate. As long as heat energy is pumped into the body, the oscillating charges continue to vibrate and emit radiation. What's more, they are capable of absorbing any amount of energy from outside and in turn emitting any amount of energy. In particular, charges which oscillate with very high frequency are capable of emitting large amounts of radiation at the high frequency (i.e. short wavelength) end of the spectrum. There is an additional factor too; just as there are vastly more numbers above any given number, there are also vastly more frequencies, at which oscillating charges can vibrate, above say the typical frequency of visible light. One result of this line of thinking was that a theoretical plot of a black-body spectrum matches experiment very well at the low frequency end of the spectrum but as we move towards the ultraviolet and beyond, the plot goes through the roof. This breakdown of the theoretical model became known as the 'ultra violet catastrophe'.

Clearly something was wrong! The German physicist Wilhelm Wien (Wien is also the German name for the Austrian capital Vienna and is pronounced 'veen') nearly got things right but his theoretical plot didn't quite match the experimental one at the longer wavelength end. It was another German physicist, Max Planck who introduced what even he himself regarded as a bit of a 'fix'. Planck assumed that the oscillating charges could not absorb or emit radiation continuously but only in whole number multiples of a fundamental unit of radiation energy equal to $h \times v$. Here v is the frequency of the oscillating charge and h is a constant now known as *Planck's constant*. What Planck was saying was that instead of absorbing and emitting radiation continuously, an oscillating charge does one of three things; it absorbs a small packet or *quantum* of energy, it emits a small quantum of energy or it simply does nothing.

At first this doesn't seem to fit in with our everyday experience of things; we switch on a lamp and radiation in the form of visible light appears to stream from the lamp's tungsten filament as a continuous beam. This though is what we see with our crude human senses. If we had the supersenses of a god, what we would see is this; the lamp's filament would contain a vast population of countless billions of charges. At any given instant some of these charges would be absorbing quanta of energy (the energy here comes from the electric current which heats up the filament), while others were emitting quanta of radiation and still others would simply be sitting there waiting to either absorb or emit. At some other instant, the scene would be similar, with previously idle charges now absorbing or emitting and so on. Zoom out to our normal crude way of seeing things and the whole thing appears as a continuous flow of radiation. Another feature of Planck's idea is that any given charge will not keep absorbing quanta ad infinitum; sooner rather than later, it will re-emit quanta which it has absorbed and this prevents infinite amounts of energy being accumulated and then radiated particularly at high frequencies. It really is as simple as that; the 'strange stuff' is only happening at a tiny subatomic level and the nature of this strange behaviour is

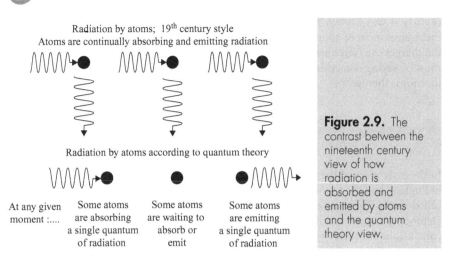

Figure 2.9. The contrast between the nineteenth century view of how radiation is absorbed and emitted by atoms and the quantum theory view.

such that when seen at a larger scale it has 'blurred out' to what we humans perceive as 'normal'.

This worked—brilliantly! The sceptical Max Planck announced his results to the equally sceptical Berlin Physical Society on 14 December 1900. This date is now seen as marking the birth of quantum theory. One final point; while Planck suggested that radiation is absorbed and emitted by a black body only in the form of quanta; he still regarded the radiation itself to be in the form of simple continuous electromagnetic waves. In 1905 however, Albert Einstein showed that to explain the photoelectric effect (the light metre in your camera uses the photoelectric effect; incoming light falls on a special sensor, which converts the energy of the light into an electric current, which can be measured and used to tell you what exposure time to use) you need to assume that light itself and indeed all forms of electromagnetic radiation exist in the form of small packets or quanta. A quantum of electromagnetic radiation is called a *photon*. There is even a direct connection between the energy of a photon and the wavelength of its associated electromagnetic wave. This comes from Planck's formula above and simply says that the energy E of a photon is given by

$$E = h \times \nu \tag{2.2}$$

And because $\nu = c/\lambda$ from Eq. (2.1)

$$E = h \times c/\lambda \tag{2.3}$$

Here, h is Planck's constant again and c is the speed of light. Notice that it also fits in with what we were saying at the beginning; that a shorter wavelength corresponds to electromagnetic radiation of higher energy. This idea, that light itself comes not so much in a continuous stream of electromagnetic waves but rather in what might be described as 'wave packets' containing a single quantum of energy proved to be of profound importance. This quantum aspect of light's dual personality will play a pivotal role in our story of spectroscopy. However, we now need to deal with the other principal player in our story—matter.

We've talked so far about radiation from a hot body resulting from oscillating electric charges situated in the outer layers of the body. Even by the end of the nineteenth century it was known that all objects are made of atoms. The oscillating charges themselves must of course form part of the structure of these atoms and the fact that the charges can only absorb and emit radiation in the form of small packets or quanta must be saying something pretty important and fundamental about the structure of the atoms themselves.

Atoms

In the early years of the nineteenth century, the chemist John Dalton from Cumbria in England laid the foundation of the theory that all matter was made of tiny particles called atoms. The term 'atom' had in fact been coined by the ancient Greek philosopher Democritus who formulated a purely philosophical theory that atoms existed. Dalton's conclusions however were based on many chemistry experiments which basically showed that a specific amount of one substance such as carbon would only react with a specific amount of another substance such as oxygen to produce carbon dioxide. Dalton's way of explaining this was that chemical elements are made of atoms and what's more atoms of different elements had different but very specific masses. Also, all atoms of a given element were identical to each other.

By the end of the nineteenth century, it was realised that atoms were some kind of mixture of positive and negative electric charge. Dominating the scene at this time was the English physicist Joseph John Thomson who realised that so-called 'cathode rays' were in fact tiny particles which carried a negative electric charge. He determined the ratio of charge to mass for these particles and concluded that they were far tinier than atoms. Thomson proposed that these cathode ray particles were in fact a component of atoms themselves and even put forward a model for the structure of atoms. Thomson's model atom consisted of a uniform sphere of positive electric charge in which were embedded just the right number of his negatively charged particles to render the atom electrically neutral. This model subsequently became known as the 'plum pudding' model. As for Thomson's negatively charged particles, they were *electrons* and Thomson was a little annoyed that it was others who coined the name, which in fact is the ancient Greek name for the fossilised tree resin amber. This had been used for spinning thread because fibres conveniently clung to it—they had in fact become electrically charged as a result of friction associated with the spinning motion.

Thomson had a student called Ernest Rutherford; a New Zealander with a reputedly brash personality. Rutherford was passionate about the recently discovered phenomenon of radioactivity and it was he who coined the terms alpha, beta and gamma radiation. Alpha radiation or alpha particles were known to be atoms of helium, which were positively charged, and during the early years of the twentieth century, Rutherford carried out a series of famous experiments in which alpha particles were 'fired' at extremely thin sheets of gold foil. The locations of emerging particles were recorded on photographic plates placed behind the foil as illustrated in Fig. 2.9. Most particles were seen to pass straight through the foil more or less unhindered which suggested to Rutherford that atoms consisted mostly of empty space. This of course was in stark contrast to Thomson's uniform spheres of charge. More striking was the fact that some particles were deflected by large angles, some even recoiling backwards.

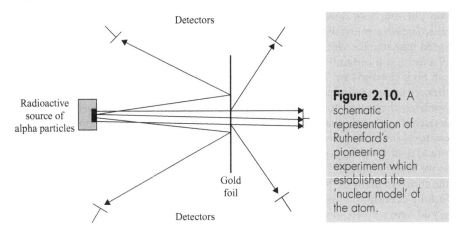

Figure 2.10. A schematic representation of Rutherford's pioneering experiment which established the 'nuclear model' of the atom.

Clearly, Rutherford was smart enough to place photographic plates all around his apparatus. This suggested to Rutherford that atoms consist of a central, positively charged nucleus which contains almost all of the atom's mass. The electrons existed in a region some distance from the nucleus. The simplest chemical element was known to be hydrogen and Rutherford suggested that its nucleus was the fundamental positively charged particle; he called it a *proton*.

This was the birth of the nuclear model of the atom which we are all familiar with today. The actual charge carried by an electron was eventually determined by the American physicist Robert Millikan in his famous 'oil drop' experiment. This enabled the mass of the electron to be determined at about 9×10^{-31} kg; this made it 1/1837 of the mass of a hydrogen atom. It was eventually realised that the position of an element in the periodic table was determined by the number of protons in the nuclei of its atoms; hydrogen has just one, helium two, lithium three, etc. and this is also known as the *atomic number*; often denoted by a capital 'Z'. This also determines the number of electrons in a neutral atom. It eventually became apparent that the mass of most atoms was much greater than could be accounted for by simply assuming that the nucleus was made entirely of protons. The deficit was made up by the discovery in 1932 of the neutron; a particle of slightly higher mass than the proton but with no electric charge. So the standard model of an atom is that of a nucleus which consists of protons and a slightly varying number of neutrons (for a given element, different numbers of neutrons result in different *isotopes*—for example, carbon 12 has six protons and six neutrons in its nucleus whereas carbon 14 has six protons and eight neutrons).

In this model then the nucleus contains almost all the mass of the atom and the electrons which match the number of protons, 'orbit' the nucleus. Note the inverted commas on the word 'orbit'. The stylised pictures that we've all seen in kids' science books, depicting an atom as some kind of miniature solar system can be misleading. Firstly, it was quickly realised that there was a problem with Rutherford's model; if an electron was some kind of particle carrying a negative electric charge, then the fact that it was in orbit around the nucleus meant that it was in fact accelerating. Most people think of acceleration as a change of speed such as what happens when you put your foot down in the car. Acceleration however also means *a change of direction*; on a roundabout ride at a fun fair, you are accelerating even though your speed may stay the same. Basically acceleration is what results from the application of a force; you certainly feel some kind of force when you take a sharp bend in your car. So if electrons are 'orbiting' the nucleus of an atom, they are doing so because they are under the influence of the electric force of attraction between them and the positively charged protons. They are therefore accelerating and as Maxwell had shown, accelerating electric charges emit electromagnetic radiation. Rutherford's 'orbiting' electrons should be constantly emitting radiation and losing energy in doing so. They would then spiral very quickly into the atom's nucleus; this obviously doesn't happen so we have a problem. The solution of this problem was quantum mechanics; it took the efforts of many of the twentieth century's greatest physicists and quite a few years to fully develop the theory. Right from the start it became clear that things which would have seemed 'obvious' and 'common sense' to a nineteenth-century physicist would have to be abandoned in order to develop a theory which correctly explained the behaviour of atoms. This behaviour included of course the nature of spectra and the patterns and observed wavelengths of spectral lines. The first thing which had to be assumed, in this case by the Danish physicist Neils Bohr in 1913, was that contrary to what Maxwell's theory said, electrons *could* exist in some kind of stable orbits in an atom; indeed the word 'orbit' became replaced by the term 'energy level'. The bottom line is; it's okay to imagine electrons as tiny particles orbiting the nucleus of an atom *provided* you don't assume that they behave like tiny pool balls.

Summary

- Light is an electromagnetic wave but it can also be thought of as a stream of particles called photons.
- The energy of a light wave or of a photon of wavelength λ is given by $h \times c/\lambda$, where c is the speed of light and h is Planck's constant.
- Visible light forms a narrow part of the electromagnetic spectrum.
- The visible light spectrum runs from about 4000 Å in the violet to about 7000 Å in the red.
- A continuous spectrum is produced by a hot dense gas–liquid or solid. The continuous part of any spectrum is referred to as the continuum.
- An emission spectrum is produced by a hot low-density gas and consists of a series of bright (emission) lines whose wavelengths depend on the chemical make-up of the gas.
- An absorption spectrum is produced when a continuous spectrum shines through cold thin gas to produce a series of dark (absorption) lines superimposed on the continuum. Again the wavelengths of the lines are determined by the chemical make-up of the intervening gas.
- A black body is an object that emits as much radiation as it absorbs at all wavelengths. It is possible for a non-black body to emit as a black body over a restricted range of wavelengths.

- The distribution of energy as a function of wavelength depends on the body's temperature and takes the form of a 'bell-shaped' curve. This shape can only be explained by assuming that radiation is absorbed and emitted in tiny packets called quanta.

- The wavelength of peak emission for a black body decreases with increasing temperature; this is called Wien's Law.

- All atoms consist of a tiny central nucleus, which contains almost all of the atom's mass.

- The nucleus consists of two types of particle; the positively charged proton and the electrically neutral neutron.

- The nucleus is 'orbited' by negatively charged electrons, which have very low mass but charge equal and opposite to that of a proton.

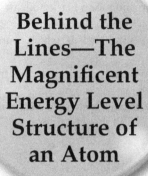

Behind the Lines—The Magnificent Energy Level Structure of an Atom

Out in the 'wild' a free electron can do what it wants and have any amount of energy that it wants. Within an atom though, things are very different; like a guy sent to jail, the electron faces a tough regime governed by set rules which restrict freedom of movement; though just as is sometimes the case with a prison, electrons can and do escape.

In this chapter we'll learn about the energy level structure of atoms and about the ways in which electrons can move around these energy levels. Sometimes an electron can leave an atom altogether while other free electrons can be captured by atoms. These are the basic processes which make spectra the way they are and by the time you reach the end of this chapter you'll understand where absorption lines and emission lines come from. You'll also learn why there is order in what at first seems like complete chaos in the jumble of lines in many spectra.

A lot of ground is covered in this chapter so take it nice and easy, one piece at a time.

Energy Levels

Nineteenth-century physics says that if an electron orbits the nucleus of an atom, it will rapidly lose energy and fall into the nucleus. Stable atoms exist however so something is wrong! It must be possible for an electron to reside in some kind of 'stable orbit' within an atom because this is what everyday experience tells us is happening. The real problem here is that no one even to this day knows exactly what an electron is, just as no one knows exactly what an electric field is. What the physicists had to do in the early decades of the twentieth century was to work out a set of principles and rules which show how these stable electron orbits are configured; this set of principles forms a large and important part of *quantum mechanics*. Quantum mechanics is able to describe in great detail and with great accuracy the behaviour of electrons in atoms without even knowing exactly what electrons are; in this sense it's actually not unlike Newtonian mechanics which can predict the orbits of planets and the motion of balls on a pool table without actually knowing that these things are made of atoms.

Unlike planets and pool balls, however, we have no 'picture' in our mind of what electrons are or might be like. In the long run it probably does no harm to think of an electron as an incredibly tiny particle which carries a negative electric charge, *provided* we don't necessarily expect it to behave like a tiny pool ball. The special stable orbits where the electrons move around the nucleus of an atom are determined by the laws of quantum mechanics not Newtonian mechanics. They are uniquely defined and are discreet and distinct from one another; an electron *must* move on one of these orbits for an atom to be stable; any kind of motion 'somewhere in between' is not allowed. Despite this, quantum mechanics can't say exactly whereabouts on its orbit an electron actually is; it can only say where it is most likely to be. So while the location and shape of the orbits are clearly defined by quantum mechanics, the uncertain motion of the electrons gives them a 'fuzzy' character. A couple of things that do hark back to Newtonian mechanics are the fact that in many cases the stable orbits are not circular. Just as planets move on orbits which are elliptical, electrons can often be thought of as doing the same; the second thing is that while some electron orbits are spherically distributed around the nucleus so there is no orbital plane as such, many orbits do have a well-defined orbital plane or orientation and this as we'll see later has a very important role to play.

The size and shape of a stable electron orbit is determined by the energy of an electron occupying the orbit; for this reason the words 'orbit' or even 'stable orbit' are not used; instead they are replaced by the term *energy level*. Every atom of every element has a set of energy levels; they are there even if they are not occupied by an electron. They often provide a location where an electron can stay happily forever if need be; very often though an electron will spend only a finite and often exceedingly short amount of time in a particular energy level before it 'decides to move on'. One thing is certain; energy levels are very exclusive, they can only be occupied by one electron at any one time. This is a very important rule from quantum mechanics called the *Pauli Exclusion principle* and it makes each energy level in an atom unique. We can think of an atom as consisting of a nucleus surrounded by a set of unique energy levels each of which may or may not be occupied by an electron; if it is occupied then the energy level itself determines the amount of energy that its occupying electron possesses. The actual energies involved are largely determined by the nucleus of the atom; in particular the number of protons (i.e. number of positive electric charges) and hence by the actual element we are dealing with. Beyond this, the nucleus of an atom plays no role in the formation of atomic spectra aside from one or two effects which are subtle in the extreme and which needn't concern the amateur spectroscopist, nor indeed most professionals. The energy levels can also be affected by the presence of other electrons within multi-electron atoms; electrons are negative electric charges and they are bound to affect each other as they 'buzz' around within the atom; more on this later.

One final thing to say here is that a free electron is said to have a 'positive' amount of energy, whereas an electron which is confined to an atom has a 'negative' amount of

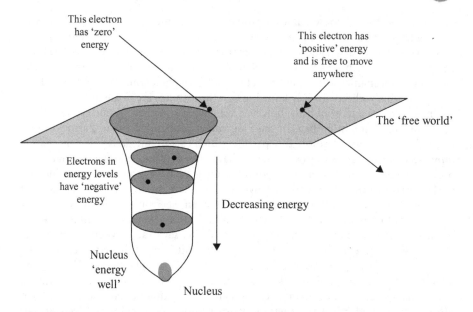

Figure 3.1. Physicists define things so that an electron which is just outside of an atom has 'zero' energy. Energy levels can then be thought of as stable zones within the energy well of the nucleus; the closer they get to the nucleus the more 'negative' their energy.

energy. There's nothing exotic about this—the electron still has energy when within an atom but it has less energy than a free electron. This is because it is held by the electric field of the atomic nucleus; think of it as a kind of 'electric field well' with the nucleus at the bottom. The further down this well the electron is, the lower its energy. Physicists define things so that an electron just 'hovering' outside of an atom has 'zero energy' and thus a free electron would in general have more energy than this 'critical' electron and an electron bound in an atom would have less. Also an electron within an atom has less energy if it is closer to the nucleus and more energy if it is further away.

Electron Energies

The standard unit of energy is called the *joule*. This is quite a large amount of energy, enough for example to move a 1 kg bag of sugar 1 m; an electron in an atom has much less energy than this. Imagine rigging up a simple electric circuit with a 1 V battery and a couple of metal plates as shown in Fig. 3.2. When we switch on the battery, an electric field exists between the metal plates and if we could place an electron next to one of the plates, it would move under the influence of the electric field to the other plate. In doing so it gains energy; the amount is very small—only 1.602×10^{-19} J. This tiny amount of energy is called one *electronvolt* or 1 eV for short. Electrons bound in atoms have (negative) energies of the order of several electronvolts.

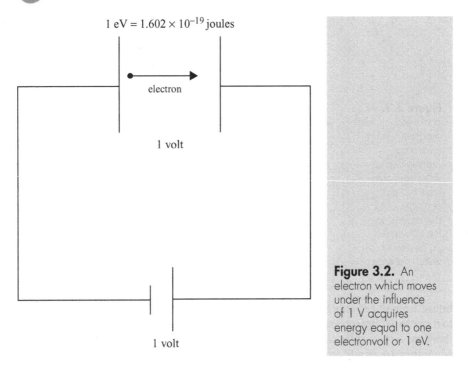

Figure 3.2. An electron which moves under the influence of 1 V acquires energy equal to one electronvolt or 1 eV.

Quantum mechanics shows that energy levels within atoms can have various three-dimensional shapes and sizes but the important thing for spectroscopy is simply the amount of energy associated with a given level. So it is much easier to depict energy levels on a diagram as horizontal lines (or little shelves) with a scale running down the left-hand side which gives the actual energies in electronvolts of the levels themselves. An energy level which is closer to the nucleus and so has less energy will be below a level of higher energy. The bottom line of this *energy level diagram* represents the nucleus of the atom and a horizontal line at the top marks the level for an electron which is just free; this of course corresponds to an energy of 0 eV. This 'freedom line' is usually shown with a hatched area above to depict the outside world of free electrons and this zone is referred to as the *continuum*. Don't however confuse this with the continuum or continuous part of a spectrum.

The final point here is that in all atoms, the energy levels get closer together as they get nearer to the continuum. This means that energy differences for adjacent levels are not so great near the periphery of an atom but can be quite large near the nucleus. The spacing of the energy levels is as you'd expect determined by quantum mechanics.

Electron Transitions

An electron in an atom can move from one energy level to another but to do this it must either acquire energy from outside the atom to move to a level of higher energy or lose energy to move to a level of lower energy. The process of an electron 'moving

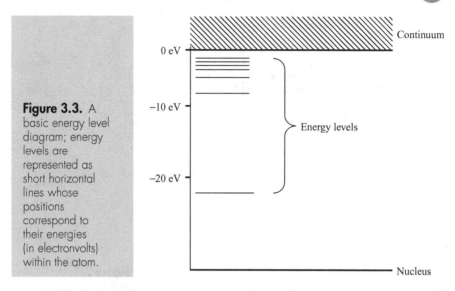

Figure 3.3. A basic energy level diagram; energy levels are represented as short horizontal lines whose positions correspond to their energies (in electronvolts) within the atom.

house' from one energy level to another is called an *electron transition*. For an electron to move to a different energy level, that level must be empty of course and what's more there are certain rules which govern the electron's 'choice' of destination levels even when they are empty (more on this shortly). Basically though an upward transition results from absorption of energy from outside and a downward transition causes loss or emission of energy. Clearly, upward electron transitions are going to play a pivotal role in the production of absorption spectra and conversely, downward transitions will result in emission line spectra.

A transition (upward or downward) in which the electron remains within the atom is often referred to as a *bound–bound* transition. There are a few other terms like this which often appear in the literature; if an electron receives sufficient energy from outside, it can escape from the atom altogether and become a 'free' electron. This process of electrons escaping from atoms is called *ionisation* and a transition which results in ionisation is called a *bound–free* transition. The free electron could at some later time be captured by another atom; this process is called *recombination* and the relevant electron transition is called a *free–bound* transition. Finally, our free electron may remain free but receive a quantity of energy from its surroundings; the electron now has more energy and so it has undergone a kind of transition called a *free–free* transition.

In a bound–bound transition, the electron loses or gains energy and the amount gained or lost is determined by the energy difference between the two energy levels involved in the transition.

Why do we need to know this? Very often (though not always, as we'll see later) the energy absorbed or emitted by an electron undergoing a transition is in the form of light; i.e. a photon and Planck's formula gives us an easy way to convert energies into wavelengths and vice versa. Remember Planck's formula states that the energy E associated with light of wavelength λ is given by;

$$E = h \times c/\lambda \tag{3.1}$$

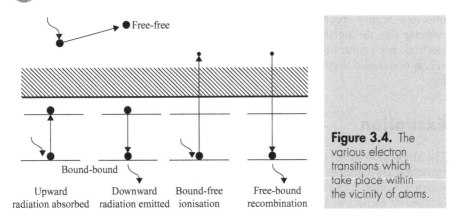

Bound-bound

Upward Downward Bound-free Free-bound
radiation absorbed radiation emitted ionisation recombination

Figure 3.4. The various electron transitions which take place within the vicinity of atoms.

Here h is Planck's constant (6.626×10^{-34} J s) and c is the speed of light (3×10^{8} m/s). To get a wavelength from a quantity of energy, we need to rearrange Planck's formula to read

$$\lambda = h \times c/E \qquad (3.2)$$

With E in electronvolts and using the above values for h and c there's a very simple formula for converting electronvolts to angstroms; this turns out to be

$$\text{wavelength in angstroms} = 1.24033 \times 10^{4}/\text{energy in electronvolts} \qquad (3.3)$$

So for example, if the energy difference between two levels in an atom of hydrogen were 1.89 eV, an electron transition between these two levels would result in either the emission or the absorption (depending on whether the electron suffered a downward or an upward transition respectively) of light at a wavelength of 1.24033×10^{4} divided by 1.89 which equals 6562.6 Å (remember Å is the symbol for angstroms). This is almost equal to 6563 Å a number which is well known to many amateur astronomers as the wavelength of the hydrogen alpha or Hα line; light which is responsible for the reddish colour of galactic nebulae like the Orion Nebula and the North American Nebula.

One final and extremely important thing to say about bound–bound transitions is this; an electron which drops to a lower energy level will lose energy and this energy will be emitted as light of a specific wavelength. For the electron to be restored to its original energy level, it must receive energy in the form of light of *exactly the same wavelength*. This might seem obvious at first but suppose our electron were to receive a bit more energy than that which it originally lost; this could be in the form of light with a slightly shorter wavelength. Would the electron be restored to its original energy level with a bit of energy to spare? The answer is no! The incoming photon may move off in a different direction; this is called *scattering* but light of any wavelength other than that which corresponds exactly to the energy involved in the transition will have no effect on the electron with one exception; if the energy of the incoming light is sufficient to remove the electron from the atom altogether, then our electron will be set free, otherwise it will stay put. Bound–bound transitions thus involve exact energies which translate into exact wavelengths which in turn translate into well-defined spectral lines. Setting electrons free on the other hand involves dealings with the 'outside world' and as long as there is sufficient energy to free the electron, any amount of energy can be used. In fact as we would expect, the more energy that is used to free the electron, the more energy it will have after it has been set free.

Excitation

The absorption of energy from outside causing an electron to undergo an upward transition to a higher energy level is called *excitation* and an atom which has undergone such a transition is said to be in an *excited state*. The opposite process, not surprisingly is called *de-excitation*. As described above, excitation by absorption of radiation is an exact process—only light of the correct wavelength will do. However, there is another way to achieve excitation; the atoms of a gas for example can be raised to an excited state by simply heating the gas to a higher temperature. This process is called *thermal excitation*; raising the temperature of the gas makes the gas atoms move around at higher speeds. This results in more frequent collisions between atoms ('collision' is a bit of a misnomer here; two atoms will actually have what might best be termed a 'close encounter'; they will exchange energy—one atom will gain energy, the other will lose and they will move apart). The result is that the collisional energy gained by some of the atoms will be just the right amount to raise them to an excited state; i.e. an electron will jump to a higher energy level. As we shall see later, thermal excitation has very important consequences for stellar spectra.

An atom which is not in an excited state, i.e. none of its electrons are in a higher than normal energy level, is said to be in *the ground state*.

Ionisation

If an electron does receive sufficient energy from outside it will escape from the atom. This process; a bound–free transition, is called *ionisation* and the atom as a result becomes *ionised* and is then known as an *ion*. The incoming energy may be in the form of a photon of sufficient energy to free an electron and this process is often referred to as *photoionisation*. As with excitation, it's also possible to ionise the atoms of a gas by sufficient heating and this process is called *thermal ionisation*. As with thermal excitation, thermal ionisation, as we shall see is of fundamental importance in the formation of stellar spectra.

The simplest of all atoms, hydrogen (H) has just one electron, so in this case ionisation can happen just once leaving a *hydrogen ion*. Element number two, helium (He) has two electrons and it can become singly ionised and possibly doubly ionised. Clearly more complex atoms can lose several electrons and become multiply ionised. Astronomers have a very specific terminology for how many times an atom has become ionised; Take for example iron whose chemical symbol is Fe; if the iron atom is still in possession of all of its 26 electrons (any atom possessing all of its electrons is referred as a *neutral atom*), then it is written as FeI; i.e. the chemical symbol followed by the Roman numeral I. Iron which is singly ionised, i.e. has lost one electron, is written as FeII and so on with FeIII, FeIV, etc. The spectra of some symbiotic stars show evidence

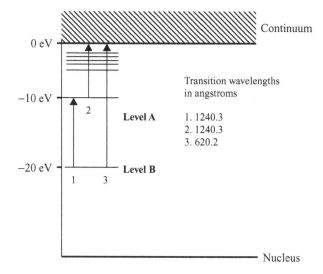

Transition wavelengths
in angstroms

1. 1240.3
2. 1240.3
3. 620.2

Figure 3.5.
Wavelengths
corresponding to
electron transitions
involving either
excitation or
ionisation can be
easily calculated
from an energy level
diagram like this by
using Eq. (3.3).

of FeXIV, i.e. iron atoms which have lost 13 or half of their normal compliment of 26 electrons. Galactic nebulae like the Orion Nebula contain large amounts of ionised hydrogen and so are often referred to as HII regions.

It clearly takes energy to ionise an atom and as with the energy involved in a bound–bound transition, this energy is usually given in electronvolts or eV. The energy needed to set free an electron from a given energy level in its parent atom is called the *ionisation potential* for that particular energy level. Again using the simple formula (3.3), it's easy to convert an ionisation potential to a wavelength and this then gives the wavelength of light which is needed to ionise an atom from that particular level. For example, the ionisation potential for hydrogen (i.e. the energy needed to remove the electron from the lowest energy level) is 13.598 eV. So if we divide 1.24033×10^4 by 13.598, we get 912.1 Å; photons at this wavelength are in the ultraviolet part of the spectrum.

In multi-electron atoms, the outer electrons are less tightly bound to the atomic nucleus. This is not just because they are further away from the nucleus but also because the inner electrons can 'screen off' much of the electric charge of the nucleus. The result is that relatively little energy is needed to remove them; hence the ionisation potentials are low. Inner electrons are generally much more tightly bound and so the ionisation potentials are correspondingly higher. If astronomers detect evidence of highly ionised multi-electron atoms in a spectrum, it's a sure sign that the object involved includes a source of high energy, i.e. very high temperatures or short wavelength radiation. The single outer electron in a sodium (Na) atom has an ionisation potential of only 5.139 eV and thus needs photons at 2413.6 Å for ionisation to take place, By contrast, it takes 47.286 eV or a 262.3 Å photon to remove one of the inner electrons.

The opposite process to ionisation is called *recombination*, where an atom captures a free electron. Recombination plays a fundamental role, as we shall see later, in the production of the emission line spectra of objects like galactic nebulae and planetary nebulae.

It All Comes Down to the (Quantum) Numbers

Every energy level in an atom is unique and is identified by its own unique code—a set of four numbers which physicists call *quantum numbers*. The numbers have names which in some cases are 'borrowed' from pre-quantum physics and you have to be a bit careful not to read too much nineteenth-century physics into these names. The four numbers are as follows:

- Principal quantum number—n.
- Angular momentum quantum number—l.
- Magnetic quantum number—m (often written m_l).
- Spin quantum number—s (sometimes written m_s).

All of these numbers are simple whole numbers, sometimes including zero; there are no fractions or decimals with the sole exception of the spin quantum number s which is always equal to either $+1/2$ or $-1/2$. The magnetic quantum number m is as you may expect very much to do with external magnetic fields and has an important role to play in certain astronomical spectra; we'll have a lot to say about m in Chapter 11. The terms angular momentum and spin come from pre-quantum physics; traditionally, angular momentum is associated with rotation or the orbital motion of say a planet. Here the angular momentum quantum number l is also associated with the shape of an energy level or the way it is distributed around the nucleus of the atom. Its real importance to the astronomer doing spectroscopy is that its value contributes to the actual energy associated with an energy level. The same thing applies to the spin quantum number s; its value affects the actual energy of a given level.

Configuring the Energy Levels

The energy level structure of any atom consists of a series of principal or main levels, each one identified by its principal quantum number n. These main levels basically tell us how close in to the nucleus the electrons are; n has the values 1, 2, 3, 4, and so on with the lower values identifying levels which are closer to the nucleus. So level 1 ($n = 1$) is closest to the nucleus, level 2 ($n = 2$) is the next furthest out and so on. Each of these principal levels is itself divided into a series of sublevels. Each of these sublevels keeps the same value for n but is identified by a different value of l. The number of 'l sublevels' depends on the value of the principal quantum number n and in fact is equal to n itself. So the $n = 1$ level is its own sublevel whereas level $n = 2$ divides into two sublevels, etc. The angular momentum quantum number l takes on the values 0, 1, 2, etc. up to the value $n–1$. So for example the $n = 3$ level divides into three sublevels identified by l values of 0, 1 and 2.

So far we've got main (n) levels and (l) sublevels. Now each sublevel (l level) itself divides into a series of sub-sublevels. Each of these is identified by the magnetic quantum number m. There is always an 'm sub-sublevel' identified by $m = 0$; further 'm levels' are then identified by values of m equal to $+1$, -1, $+2$, -2, etc. up to $+l$ and $–l$. So for example, the $l = 2$ sublevel divides into five 'm sub-sublevels' with

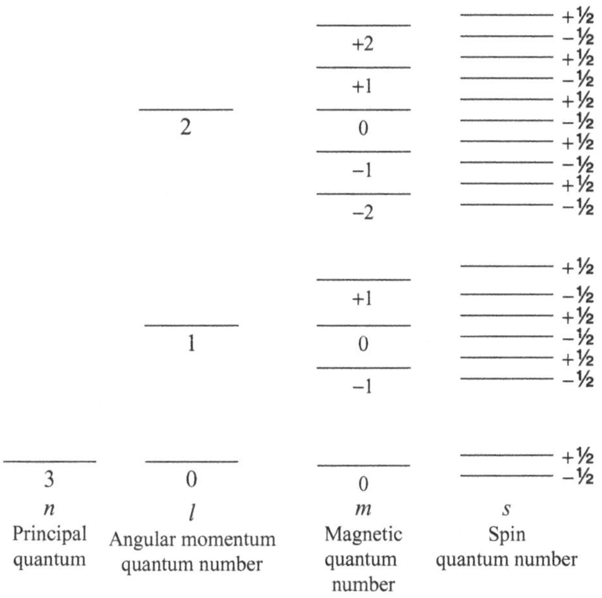

Figure 3.6. Subdivision of the $n = 3$ principal energy level.

n Principal quantum	l Angular momentum quantum number	m Magnetic quantum number	s Spin quantum number

values of m equal to $-2, -1, 0, +1, +2$. Again, if l equals 0, it is its own single m level with m also equal to 0.

Finally, each m sub-sublevel divides into two levels identified by the spin quantum number s with s having the values $+\frac{1}{2}$ and $-\frac{1}{2}$. Fig. 3.6 shows how this works for the $n = 3$ level.

As mentioned above the principal quantum number' n in effect tells us how close in to the nucleus the level is; higher values of n correspond to levels which are further out. The different l sublevels into which it divides tell us basically about the way the level is distributed around the nucleus, i.e. its three-dimensional shape; the sublevel with the highest value of l (i.e. when $l = n - 1$) is symmetrical about the nucleus, i.e. spherical. This means that an electron in this level stays the same distance from the nucleus. Lower values of l correspond to sublevels which are increasingly asymmetrical and this causes an electron in such a level to spend more time closer to the nucleus. The result is that lower l values correspond to levels of slightly lower energy; this has the odd consequence that an electron in a sublevel with a higher value of n but with a low value of l can actually get closer to the nucleus (and hence have lower energy) than an electron in the next n level down, provided that electron has a high value for l.

Each l sublevel divides into a series of sub-sublevels identified by m. Under normal circumstances, these 'm sub-sublevels' all have the same energy for a given value of l. Because they have the same energy, physicists call these sub-sublevels *degenerate*—they are rather like dormant levels which are allowed to contain one electron each but which are otherwise indistinguishable from one another. They become important when the atom finds itself in the presence of an external magnetic field; the m sub-sublevels then split off from each other, each having a slightly different energy to the others.

We'll look at the effect of this in Chapter 11. The subdivision of the atomic energy levels ends with an m sub-sublevel splitting into two, one with spin quantum number s equal to $+\frac{1}{2}$ and the other $-\frac{1}{2}$. An electron with $s = +\frac{1}{2}$ is often referred to as being 'spin up', the other being referred to as 'spin down'.

There is one very important exception to all of this; hydrogen, the simplest element has m levels which are degenerate just as with atoms of other elements. However for hydrogen, the l levels are also degenerate; in other words the only things which determine the energy of an electron in a hydrogen atom are the principal quantum number n and the spin quantum number s.

So this is how the energy levels in every atom are built up; the actual energy associated with each level depends largely on the number of protons in the nucleus, i.e. on which element we are actually dealing with. For atoms near the beginning of the periodic table, like hydrogen, helium, etc., the energy levels are for the most part empty but as we move through the periodic table, the energy levels progressively get filled with electrons. From what we have learned above, it's easy to see how many electrons it takes to fill each principal energy level. Take $n = 1$ first; l has only one value, namely 0 and so does m; s can have a value of either $+\frac{1}{2}$ or $-\frac{1}{2}$. So for the $n = 1$ level only two different sets of quantum numbers are possible; $1, 0, 0, +\frac{1}{2}$ and $1, 0, 0, -\frac{1}{2}$ and therefore this level can only accommodate two electrons.

For the $n = 2$ level, l can have values of 0 and 1; let's take $l = 0$ first. In this case m again only has the value 0 but s can have values $+\frac{1}{2}$ and $-\frac{1}{2}$, so there are two electrons in the $l = 0$ sublevel. For $l = 1$, m can now have values of $-1, 0$ and $+1$; that makes three values and each of these can have a value of s again equal to $+\frac{1}{2}$ or $-\frac{1}{2}$ making a total of six electrons all with l values equal to 1. Adding this to the two electrons with l equal to 0, gives a grand total of eight electrons necessary to fill the $n = 2$ level. The quantum numbers work out as follows:

$n = 2$	$l = 0$	$m = 0$	$s = +\frac{1}{2}$	
			$s = -\frac{1}{2}$	
	$l = 1$	$m = -1$	$s = +\frac{1}{2}$	
			$s = -\frac{1}{2}$	8 electrons in all
		$m = 0$	$s = +\frac{1}{2}$	
			$s = -\frac{1}{2}$	
		$m = 1$	$s = +\frac{1}{2}$	
			$s + -\frac{1}{2}$	

An atom with the $n = 1$ and $n = 2$ levels filled is an atom of the inert gas which is used for many of the advertising displays on our city streets—neon (Ne). In fact, when an atom has a principal level filled with its compliment of electrons, it is said to have a *closed shell*. See if you can work out how many electrons are needed to fill the $n = 3$ level; the answer is 18 (*hint*, look at Fig. 3.5).

Finally, there is another conventional way of labelling the l sublevels; alongside the values 0, 1, 2, 3, 4 to identify the l levels, the letters s, p, d, f, g are used. Thus, a level with angular momentum quantum number l equal to 0, is often referred to an 's level'; be careful here not to confuse this with the spin quantum number s. In turn an $l = 1$ level is called a 'p level' and so on. The origin of the letters s, p, d, f comes from early investigations of the spectrum of sodium (Na) and is not at all logical; One advantage though is that instead of talking about say, the '$n = 2, l = 1$' level, we can simply say the '2p' level or for the '$n = 3, l = 0$' level we can say the '3s' level and so on. This way of talking about energy levels is in fact standard practice.

The Rules of the Game—Selection Rules

I said at the beginning that within an atom, an electron is subject to a tough regime of strict rules which govern the way it can move around the energy levels. Of course in heavier atoms (i.e. those which are further up in the periodic table), many of the energy levels will be filled possibly more or less permanently so this in itself will inhibit which levels a restless electron can go to. However, even if our electron lives in a hydrogen atom it is still restricted in its choice of movement between energy levels. There is a set of rules called *selection rules* which come from the laws of quantum mechanics which tell our electron what it can and can't do. Here's the twist though; quantum mechanics itself doesn't actually lay down totally strict rules; it says that under certain conditions in the subatomic world, some things are much more likely to happen than others so in laying down the selection rules for electron transitions within atoms, it's really just saying that some bound–bound transitions are very likely to happen whereas others are very unlikely to happen. This is a set of rules which is basically asking to be broken and out there in the real Universe they do indeed get broken.

The selection rules for electron transitions between energy levels are as follows:

- The l quantum number must change by $+1$ or by -1.
- The spin quantum number s does not change.

The second rule says that an electron which starts off as spin up will remain a spin up electron after the transition. It means for example that if our electron is spin up and a possible transition would take it to a level whose l number differs by 1 (so the first rule is obeyed) but that level already contains a spin up electron, then it can't go there. Another example would be a hydrogen atom with its electron in the 2s level ($n = 2$, $l = 0$); this electron cannot drop down to the 1s level because it would violate the first rule. An electron in this level is in a sense stuck and levels which are like this are called *'metastable levels'*; it can only 'escape' by being excited to a higher level from which it could then drop to the 1s level—or it could break the rules!

So the selection rules are about bound–bound electron transitions which are proba-ble and those which are (quite possibly highly) improbable. Probable transitions obey the selection rules and are called *permitted transitions* and improbable ones which break the selection rules are called *forbidden transitions*. If radiation (light) is emit-ted as a result of a transition then it is referred to as either *permitted radiation* or *forbidden radiation* depending on whether the transition is permitted or forbidden. The selection rules do get broken and forbidden radiation is observed in astronomical spectra.

Order from Chaos—Spectral Series

A typical spectrum of a star consists of a continuous background called perhaps not surprisingly the *continuum* (be careful not to confuse this with the 'continuum' which represents the world of free electrons outside of an atom). In photographs this appears as a rainbow coloured strip (a shaded grey strip if it's an old photograph) and in a more modern spectrum plot as a more or less smooth curve. The photographic spectrum will be crossed by perhaps many dark lines called *absorption lines* which again in a plot of the spectrum appear as sharp dips in the continuum. These absorption lines are the result of upward bound–bound transitions in atoms which lie between us and the source of the continuum. The bound–bound transitions involve light of specific wavelengths which is why the absorption lines themselves have specific wavelengths and of course these wavelengths tell us what amount of energy is involved in the transition. Most importantly of all perhaps is the fact that lines from any element organise themselves (according to the rules of quantum mechanics of course) into individual groups called *spectral series* or also *transition series*.

Hydrogen

With hydrogen we have only one electron to worry about, so let's worry about this first and the first thing to think about is which energy level the electron starts from. Logic suggests that we start with the electron in the ground state, i.e. the $n = 1$ level. From here the electron can go to level 2, level 3, level 4, etc. Each of these transitions is the result of absorbing photons of a very specific energy, i.e. wavelength, and together they make what is known as a *transition series*. This particular transition series is called the *Lyman series* and the photons which are needed to produce it are of pretty high energy in the ultraviolet part of the spectrum. So we don't get to see the Lyman series of lines in the optical part of the spectrum. Alternatively, the same transition series can result from downward transitions which end on level $n = 1$; this produces the Lyman series of emission lines.

From the $n = 2$ level, upward transitions take the electron to level $n = 3, 4, 5$, etc. and this give us a series of absorption lines in the optical part of the spectrum called the *Balmer series*. This series of lines is so famous that just as with bright stars, the lines of the Balmer series have names, though these are not very exciting. The transition from $n = 2$ to $n = 3$ gives us the Hα line which is our old friend at λ6563. The transition $n = 2$ to $n = 4$ gives us the Hβ line and so on. Upward transitions from the $n = 3$ level give a series called the Paschen series in the infrared. Fig. 3.7 shows the energy level diagram for hydrogen; the transition series are shown by arrowed lines running between the relevant energy levels—upward pointing arrows are used for absorption lines and downward pointing arrows are used for emission lines. More advanced books and research papers etc. will very often give actual wavelengths alongside the transition arrows or actual energies in electronvolts for the various energy levels on a diagram like this, so it's very straightforward to use these values to calculate what the wavelength corresponding to a transition will be.

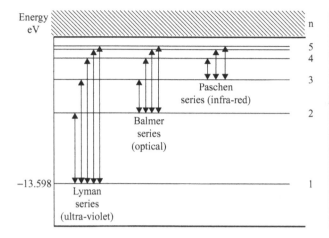

Figure 3.7. Three transition series for hydrogen; the distinguishing feature is the energy level on which transitions start and end; $n = 1$ for the Lyman series; $n = 2$ for the Balmer series, etc.

Spectral lines arrange themselves into series like the Lyman and Balmer series for hydrogen. The lines of any series always crowd together towards the short wavelength end as shown in Fig. 3.8; this is a direct result of the closer spacing of the energy levels towards the periphery of atoms. As we might expect, lines from more complex atoms are more numerous and hence more difficult to sort out but just as with hydrogen, they too arrange themselves into transition series.

Sodium and Friends

Here we go from element number 1 to element number 11 in the periodic table. Why not go to element number 2; helium? you might ask. We'll see shortly that helium has a more complicated spectrum than sodium which is why we're doing sodium first. Sodium has one thing in common with hydrogen; it has what chemists call one *valence electron*. This is also the case for sodium's chemical cousins such as potassium and caesium, etc. This means that so long as we're dealing with bound–bound transitions within the neutral atom, there is only this one valence electron to worry about. The valence electron is relatively loosely bound to the nucleus in level $n = 3$ whereas the remaining 10 electrons are bound much more tightly in closed shells in levels 1 and 2 and so don't play any part in forming the spectrum.

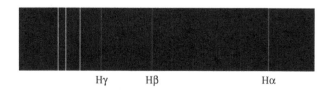

Hγ   Hβ   Hα

Figure 3.8. The lines in all spectral series crowd together towards the short wavelength end, as shown here for the Balmer series; a direct result of the closer spacing of energy levels towards the periphery of an atom.

Sodium and the other *alkali metals* as they are called do however have one big difference from hydrogen. In hydrogen remember the *l* sublevels; s, p, d, f, etc. are degenerate which means they all have the same energy. With sodium this is not the case; s levels have different energy to p levels and so on, even for the same value of *n*. This means for example that instead of just one spectral series in the optical part of the spectrum like the Balmer series for hydrogen, there are four. In fact, the letters 's, p, d and f' come from the first studies of these transition series in the optical spectrum of sodium. The letters stand respectively for 'sharp, principal, diffuse and fundamental'—names which basically don't mean anything.

Fig. 3.9 shows the energy level diagram for neutral sodium; energy level diagrams are also often called *Grotrian* diagrams after W. Grotrian who published a collection of such diagrams for simple spectra in 1928; they are also sometimes called *term diagrams*. In a more complicated diagram like this the separate *l* levels are spread out horizontally and labelled s, p, d and f accordingly and I've also included the principal quantum numbers to remind ourselves which *n* levels they belong to.

The bottom two levels which contain the tightly bound electrons are not shown. The energy scale down the left-hand side is very important here because it shows us something very interesting besides the energy values for each level. You can see for example that the 3d level is actually above the 4s level despite the 4s level having a

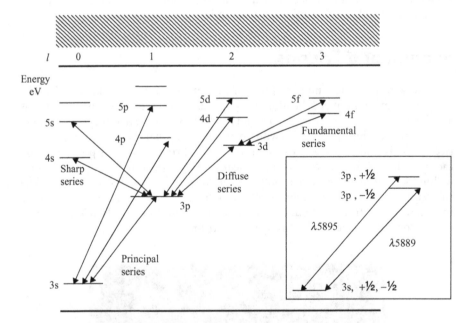

Figure 3.9. The four optical transition series for sodium (Na); notice that for each of these series, the quantum mechanical selection rule for *l* is obeyed. The inset shows the effective doubling of the 3p level due to the fact that if the electron is 'spin up' it has slightly more energy than it has if it's 'spin down'. Higher levels with *l* greater than 0 are also doubled in this way.

higher value for its principal quantum number. As described earlier, levels with a low angular momentum quantum number l have an asymmetric distribution around the nucleus and this results in the electron spending more of its time in closer proximity to the nucleus; in consequence this electron has less energy than it might otherwise have. By contrast, the 3d level has a symmetrical distribution which keeps the electron at a constant distance from the nucleus; the electron has what you could call its rightful quota of energy which turns out to be more than that for the 4s electron.

For the possible transitions, the selection rules apply; we're only dealing with one electron so whether it be spin up or spin down, its spin quantum number will remain the same. The 'l rule' as we can call it means that in any transition its value must change by 1 and only 1. This means that an electron cannot for example go from one s level to another or from an s level to a d level. Transitions take place between adjacent columns in our term diagram; we've also labelled each set of lines to indicate which series it belongs to.

The most important line for astronomers is the one which results from the transition 3s to 3p in the principal series which is seen in the yellow part of the spectrum at around $\lambda 5892$. In a low-resolution spectrum this appears as a single line but at high resolution it becomes double with its two components at $\lambda 5889$ and $\lambda 5895$. The reason for this is to do with the electron's spin quantum number s; it can either be $+1/2$ or $-1/2$. An electron in the 3p level which is spin up has slightly more energy than one which is spin down. Down in the 3s level (and this applies to all s levels in the atoms of all elements) the spins are degenerate; they're still there but for an s level, spin up and spin down correspond to equal energy. In the 3s to 3p transition a slightly shorter wavelength photon is absorbed by a spin up electron than a spin down electron. In a large population of sodium atoms some of the electrons are spin up and some are spin down, so we get two lines close together. The lines are called the 'D (as named by Fraunhofer) lines'. They are perhaps the most famous example of what's called a *doublet* line.

The single valence electron in the alkali metals such as potassium (K), rubidium (Rb) and Caesium (Cs) can either be spin up or spin down and this results in the energy levels outside of the inner core of electrons being doubled with the important exception of the s levels. Hydrogen itself has doublet levels but their separation is very small indeed. In fact, the separation in electronvolts (this of course translates into separation in wavelength) gets bigger for elements further up the periodic table; so for example lines from potassium are split wider than those for sodium. Also the splitting is wider for transitions involving low l values; hence the sodium D lines are wider than those of the diffuse series lines which involve transitions between p and d levels.

Dancing Electrons—It Takes Two (or More) to Tango

For atoms like sodium, the single valence electron has all the action and it isn't affected by the inner electrons which form tightly bound closed shells. However, if outside of the closed shells there are two or more valence electrons then these valence electrons have to share the action; because they are electric charges they interact with each other. Electrons also have 'spin'; it's called spin because it's exactly the amount of energy they would have if they were tiny spinning electric charges. A spinning electric charge generates a tiny magnetic field and these spin magnetic fields also interact. The overall result is that the energy levels for these valence electrons get modified by their own interacting electric and magnetic fields. In quantum mechanics this interaction process is called *L–S coupling* or sometimes *Russell–Saunders coupling*. Let's see how it works.

Remember when we talk about l quantum numbers and s quantum numbers, we're really talking about units of energy which electrons have over and above that which is specified by the principal quantum number n. When two or more electrons in an atom interact, these units of energy can combine to effectively add together or compliment one another or they may offset each other or even effectively cancel one another out. Even for a single electron, its total energy is enhanced if it is 'spin up', i.e. $s = +\frac{1}{2}$ and diminished if $s = -\frac{1}{2}$. A simple way to represent this is to simply add the spin quantum number to the angular momentum quantum number and call the result ' j '. So $j = l + \frac{1}{2}$ or $l - \frac{1}{2}$ and j can be thought of as representing the electron's total energy; j is often called the '*inner quantum number*'. The exception is when l equals 0; in this case the spins are degenerate in the sense that they have no effect on the electron's total energy so $j = 0$ too.

Two or more electrons will try to push each other further apart because they are after all negative charges and like charges repel. This effect is greater for higher values of l because higher l values mean that the energy levels are nearer to being spherically symmetric and this results in the electrons interacting with each other more frequently. The l values for the electrons are thus said to be *coupled*.

The electron spins are also coupled but here it is the tiny magnetic fields generated by the spins which do the interacting. Think of an electron as a tiny bar magnet; if the spin equals $+\frac{1}{2}$ the 'north pole' points up; if s equals $-\frac{1}{2}$ it points down. Several electrons which are all spin up will couple strongly and enhance each other's total

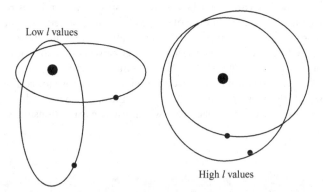

Low l values

High l values

Figure 3.10. The electrons on the left are in low l value sublevels; these are asymmetric and the electrons spend less time in close proximity to each other. The electrons on the right are in high l value sublevels which are more symmetrical; the electrons spend more time in close proximity and so there is a stronger interaction between the levels.

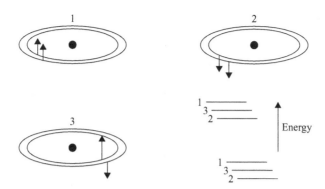

Figure 3.11. Interacting electron spins for a two valence electron atom; in case 1 the spin magnetic fields combine to enhance each other's energy level. In case 2, they combine to diminish each other's energy level. In case 3 the spin magnetic fields effectively cancel and have no effect on the energy levels. The overall result is that energy levels for two valence electron atoms in which both electrons are optically active are split into three and give rise to triplet series of spectral lines.

energy; oppositely paired spins will effectively cancel each other so there is no overall enhancement of the energy levels. Several spin down electrons will couple together to effectively diminish each other's total energy.

For a two valence electron atom, this has the effect of splitting the energy levels into three as shown in Fig. 3.11. Let's see how it works for an atom which has three valence electrons; a familiar example would be aluminium (Al). For aluminium as with sodium the $n = 1$ and $n = 2$ levels are closed shells so for the three valence electrons, life begins in the $n = 3$ level. Consider now the different combinations of spins which these three electrons can have:

Case 1			Case 2			Case 3			Case 4		
$+1/2 + 1/2 + 1/2$			$-1/2 - 1/2 - 1/2$			$+1/2 + 1/2 - 1/2$			$-1/2 - 1/2 + 1/2$		
A	B	C	A	B	C	A	B	C	A	B	C

I've labelled the electrons A, B and C; let A's angular momentum quantum number equal l (but not 0) and consider what happens to A's total energy in terms of its inner quantum number j:

- Case 1: $s = +1/2$ but s also equals $+1/2$ for B and C; the strong coupling between the electron spins effectively boosts A's 'j' value to $l + 3/2$;
- Case 2: $s = -1/2$ which offsets A's l value and this is further diminished by $s = -1/2$ for B and C; the result is $j = l - 3/2$;
- Case 3: $s = +1/2$; this gives $j = l + 1/2$; the opposite spins of B and C cancel each other and have no effect on A;
- Case 4: $s = -1/2$; this gives $j = l - 1/2$ and as with Case 3 the opposite spins of B and C have no effect on A.

Whatever level electron A finds itself in (provided it's not an $l = 0$ level), it can have any one of four possible values of total energy depending on its spin and that of its companions. The energy levels are thus split into four; and this means that electron transitions between these levels are split into four; these transitions are called *quartets*.

In the above scenario, all three electrons take part in the action and they are all interchangeable; the spin combinations have the same effect on electrons B and C. However, there is another possibility; this confines electrons B and C to their lowest possible level and for aluminium this would be the 3s level where their spins would be opposite. This then leaves electron A with all the action; electrons B and C take no part in transitions but A can be either spin up or spin down so just as with sodium the energy levels are doubled. Three valence electron atoms then, produce separate sets of doublet and quartet transitions. There's another feature here too; in the doublet scenario, electrons B and C are acting almost like a mini closed shell and help to further screen off the electric field of the atomic nucleus. This means that doublet levels are not so tightly bound; by contrast in the quartet scenario, all three electrons can move around the energy levels so the screening effect isn't there. The result is that quartet levels are closer to the nucleus and have total energies a bit lower than corresponding doublet levels.

L–S coupling causes energy levels in multi-electrons atoms to split and the degree of splitting is determined by the various possible electron spin combinations. Helium for example is a two electron atom; if both electrons are involved in transitions the combination of spins mean that each energy level is split into three (aside from the s levels) and the result is a triplet series of transitions. If as with aluminium we confine one electron to the lowest level the remaining electron has all the action but its spin must be opposite to that of the confined electron. The resulting energy levels remain single and a singlet series of helium lines results. In actual fact the splitting of energy levels in helium is extremely slight and very high-resolution spectra would be needed to show it.

The actual number of levels into which an initially single level is split is called the *multiplicity* of the level and it is simply equal to the number of active valence electrons plus one; so aluminium with its three valence electrons gives us $3 + 1$ which equals 4 for the quartet lines and $1 + 1$, i.e. 2 for the doublet lines.

Now for the usual final point; in a vast population of aluminium atoms there will be doublet transitions and quartet transitions but for the most part 'never the twain shall meet'. Transitions between doublet levels and quartet levels or indeed between any levels of different multiplicity within an atom are forbidden by quantum mechanics but not totally; when they do occur the resulting lines are called *intercombination lines*.

Ions

The spectral series formed by ions (i.e. atoms which have lost one or more electrons) are totally different from those of the neutral atom. This though turns out not to be quite so bad because when an atom loses a valence electron, it effectively becomes simply an atom with one less valence electron. For example, calcium when neutral has two valence electrons and so produces spectral series which are singlets or triplets. Ionised calcium (CaII) has just one valence electron and so produces series which are doublets like sodium. The two strongest lines in the spectrum of the Sun are Fraunhofer's H and K lines which are produced by ionised calcium. The problem can be made worse though, because the atmosphere of a star may well contain atoms of some element in both neutral and ionised form and so the spectrum contains series from both types of atom.

All in all spectra can be pretty complicated things and this is something which astronomers have to learn to cope with. The ordering of spectral series into singlets, doublets, etc. (the general term is *multiplets*) does clearly help and there is also the added 'blessing' that many complex atoms are so rare that they are not likely to be seen in stellar spectra. Even so the beginning amateur spectroscopist has a long hard road ahead of him/her.

A Final but Very Important Note

As astronomers, we're more familiar with and probably feel more comfortable with spectra which plot intensity of radiation against wavelength rather than frequency; mainly because the numbers are easier to remember and deal with. A physicist though might well prefer to plot intensity against frequency; the reason is that energy (for some transition say) is directly proportional to frequency (of radiation absorbed or emitted as a result of the transition). In other words doubling the energy means doubling the frequency; tripling the energy means tripling the frequency and so on. This means for example that transitions between a series of energy levels which were equally spaced in energy would result in a series of spectral lines which were equally spaced in frequency and this would show as a set of equally spaced lines in a spectrum but only if the spectrum plotted intensity against frequency. The same spectrum plotted using wavelength would not show equally spaced lines but lines which got closer together as the wavelength got shorter. This is because frequency is proportional to the reciprocal of the wavelength or one divided by the wavelength.

Summary

- Electrons move around the nuclei of atoms in stable orbits called *energy* levels.
- Each energy level is unique and can only be occupied by one electron at a time—a consequence of the Pauli Exclusion Principle of quantum mechanics.
- The four main quantum numbers n, l, m and s form a kind of address system for energy levels.
- An electron transition takes place when an electron moves from one energy level to another as a result of receiving energy from or losing energy to the outside.
- Transitions can be bound–bound, bound–free (ionisation), free–bound (recombination) or free–free.
- Bound–bound transitions are subject to the quantum mechanical selection rules; obeyed rules give permitted transitions; broken rules give forbidden transitions.

- Upward bound–bound transitions are referred to as excitation; downward transitions as de-excitation.
- Excitation and ionisation can be caused by raising temperatures—thermal excitation and ionisation.
- The continuous background of a spectrum is called the continuum as is the world of free electrons outside of an atom.
- Spectral lines group together in series; e.g. the Balmer series for hydrogen; the s, p, d and f series for sodium.
- Interaction between the outer or optically active electrons in multi-electron atoms causes energy levels to be split into doublets, triplets, etc.

Our Old Friend the Doppler Effect

You'd be hard put these days to pick up any book on astronomy and not find somewhere in its pages mention of the Doppler effect; this might be to do with the proper motions of stars, the radial velocities of a binary star components or perhaps most famous of all the expansion of the Universe (of course we know now that this actually isn't a Doppler effect at all but the expansion of spacetime itself, even though mathematically the two effects are identical). As we shall see in the next chapter this most valuable of effects plays a big role in spectra too; first though for completeness I shall explain it here and we'll also have a look at how important the 'clever astronomer's' relativistic version of the Doppler effect really is.

Waves and Movement

The Doppler effect is something which affects the frequency and thus the wavelength of any form of wave motion; this means it can affect water waves, sound waves and of course light waves. It occurs when there is relative movement between a source of waves and the observer who of course we assume is a fully kitted out physicist with the gadgets necessary to measure wavelengths, velocities, etc. In order for the effect to make its presence felt, the relative motion must be along the line joining the observer and the wave source or in other words directly towards or away from the observer. Motion which is directed along this line is called *radial motion* and the velocity of this motion is in turn called the *radial velocity*. If the wave source is moving at right angles, i.e. across the observer's line of sight or indeed if the observer is moving with the wave source and so is *not* either approaching it or receding from it, then there is no effect. The final thing is that it doesn't matter whether it is the wave source or the observer or both who are moving; only the relative radial velocity between the two matters and obviously it goes without saying that when we observe the Doppler effect in any astronomical situation both are moving.

How It Works

Suppose we are moving alongside a source of waves so that as far as we the observer are concerned, the wave source is stationary. We can observe or at least detect that the waves from the source are spreading out in all directions and that the wave crests (the zones of maximum electric field strength for example for a light wave) are equally spaced because they reach us at equal intervals of time. This gives the wave a well-defined and fixed wavelength. In two dimensions the wave crests will form a series of expanding concentric circles around the wave source and these are called *wavefronts*. Now let's get crazy and move towards the 'eye of the storm'; rather than sitting and waiting for the wave crests to come along; like an impatient surfer we rush headlong towards the approaching wave crest and the next one after that and on and on. The time between successive wave crests will now obviously be shorter; it's exactly the same effect as if a more laid back surfer with super powers had somehow shortened the distance between the wavefronts and this is how we observe it. By moving towards a wave source we 'see' a shorter wavelength. For a light wave this translates into seeing a bluer wave, a blue shift; the faster we head towards the wave source or the faster it heads towards us or both, the shorter the wavelength and the greater the blue shift.

Now let's turn tail and race away from the approaching wavefront; this time the wavefront has to catch us up and if the wave is moving faster than we are this will eventually happen but it now appears as if there is a longer gap between wavefronts. We see this as an increased wavelength or a red shift for light. Some waves we could outrun and of course a jet airplane which breaks the sound barrier outruns its own jet engine's sound waves; with light though we can't do this. In all of this it's the relative radial velocity which determines the wavelength shift; blue or red and the formula for this shift is very simple indeed and of course extremely useful—so useful in fact that you're likely to be using it all the time when analysing your spectra. Here it is

$$\Delta\lambda/\lambda = v/c \qquad (4.1)$$

Here $\Delta\lambda$, which is pronounced 'delta lambda', is the change in wavelength caused by relative motion between the wave source and the observer, λ is the wavelength as would be observed if the source were stationary, v is the relative radial velocity and c is the speed of light. The important thing when using this equation is to make sure that $\Delta\lambda$ and λ are in the same units, e.g. angstroms, and that v and c are also in the same units, e.g. kilometres per second. To rearrange the equation to calculate a wavelength shift for a given radial velocity, we would put

$$\Delta\lambda = \lambda \times v/c \qquad (4.2)$$

An excellent way to get a better feel for any equation like this is to try it out on a real situation and see what kind of numbers we get from it. Let's take a good round number like 100 km/s for the radial velocity and a familiar spectral line like the Hα

line at λ6563; plugging these numbers into Eq. (4.2) we get

$$\Delta\lambda = 6563 \times 100/3 \times 10^5 \text{ which equals } 2.19\,\text{\AA}.$$

So a radial velocity of 100 km/s produces a red or blue shift of about 2 Å at λ6563; shorter wavelengths will result in smaller Doppler shifts. Of course as an observational spectroscopist you're more likely to want to do things the other way round; i.e. calculate a radial velocity from what you believe to be a shifted wavelength on your spectrum. Provided you can identify the line and know its laboratory wavelength you can simply rearrange Eq. (4.1) to read

$$v = c \times \Delta\lambda/\lambda \tag{4.3}$$

So for example, a wavelength shift of 5 Å for the Hα line corresponds to a relative radial velocity of $3 \times 10^5 \times 5/6563$ which equals 228.55 km/s. At 1000 km/s the Doppler shift for Hα increases to almost 22 Å; however as I'm sure many of you know, for very high radial velocities, there is a more complicated Doppler shift equation which involves the special theory of relativity so let's have at look at this.

The Relativistic Doppler Shift

First of all, here's the relativistic Doppler shift formula

$$\Delta\lambda = \lambda \times \frac{1 + \frac{v}{c}}{\sqrt{1 + \frac{v^2}{c^2}}} - 1 \tag{4.4}$$

This is a much more involved equation than (4.2) and much greater care has to be taken if you're going to use it; best of all write a simple computer program which allows you to enter a rest wavelength and a radial velocity and then does the calculation for you. The same rule applies in that $\Delta\lambda$ and λ must be in the same units as must v and c. If we try this equation out on a radial velocity of 1000 km/s we get $\Delta\lambda$ equal to 21.91 Å compared to 21.88 Å if we use Eq. (4.2); again I've taken the Hα line at λ6563 as the rest wavelength. Clearly there isn't much difference but to explore further I've computed $\Delta\lambda$ using both Eqs. (4.2) and (4.3) for a range of radial velocities and these are summarised in Table 4.1.

Even at a radial velocity of 5000 km/s the difference in wavelength shift between the non-relativistic and relativistic cases is less than 1 Å. For the majority of situations you're pretty well okay to use the simpler formula, i.e. Eq. (4.2). One final point about using equations (4.2) and (4.4); a relative radial velocity which increases the separation of the observer and the wave source is entered as a positive number and this will give a wavelength shift $\Delta\lambda$ which is also positive. Hence we have an increased wavelength or a red shift. Relative radial velocities which decrease the separation of source and observer should be entered as a negative number and this will result in a negative wavelength shift, i.e. a blue shift.

Table 4.1 The Doppler shift $\Delta\lambda$ for the Hα line (λ6563) is shown here using both the relativistic and non-relativistic formulae for a range of radial velocities. As can be seen, for velocities up to about 5000 km/s, it's probably okay to use the simpler non-relativistic formula.

Radial velocity (km/s)	$\Delta\lambda$ Non-relativistic (Å)	$\Delta\lambda$ Relativistic (Å)	Difference (Å)	% Difference
1000	21.88	21.91	0.03	0.14
2000	43.75	43.90	0.15	0.34
3000	65.63	65.96	0.33	0.50
4000	87.51	88.10	0.59	0.67
5000	109.38	110.31	0.93	0.85
10000	218.77	222.54	3.77	1.72
50000	1093.83	1202.45	108.62	9.93
100000	2187.67	2718.48	530.81	24.26

A Very Important Point

Out there in the big wide Universe, all manner of things are moving in all kinds of directions; these could be galaxies, stars or countless atoms within the photosphere of a single star. The light that we observe these things by is more often than not going to be either red or blue shifted by the Doppler effect; but also, more often than not a source of light waves won't be moving either directly towards or away from us or exactly across our line of sight. Things out there will more generally be moving at some angle to our line of sight; either more or less towards us or more or less away from us. The actual motion of a light source can be split into two components; one part is directed across our line of sight and is called the *tangential motion*; the other component is that part of the object's motion which would carry it directly towards or away from us and this is of course called the *radial motion*. The real actual motion of the wave source through space is called perhaps not surprisingly the *space motion*.

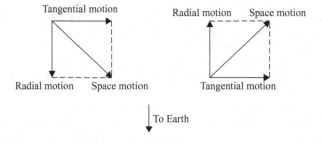

Figure 4.1. The space motion of any astronomical object can be 'split' into two components at right angles to each other: one is directed across the plane of the sky and is called the tangential motion; the other is directed either directly towards or directly away from us and is called the radial motion. The value of the radial motion both in terms of speed and direction determines the Doppler shift.

A Simple Application—Spectroscopic Binaries

One of the things we learn about in elementary astronomy books is that there are in the sky many double stars or *binaries*. Binaries are stars which are physically bound to each other by their gravitational attraction and as a result they each orbit about the common centre of gravity of the system. Provided the plane of the orbit is not face on or at right angles to our line of sight, then clearly as the two stars revolve about one another, one will generally be moving towards us while the other moves away. There will also be times when both stars are crossing our line of sight in opposite directions.

The components of many binaries are too close together to be seen separately in a telescope; in this case the spectra of the two stars blend into one combined spectrum. Due to the orbital motion however the lines in one of the spectra will periodically show a varying red shift while those of the other will show a blue shift; the lines split and move apart. As the two stars cross our line of sight the lines close up, swap shifts and move apart again. The binary nature of the system has been given away by the periodic behaviour of its spectrum; binaries which do this are called *spectroscopic binaries*.

Clearly the effect will be greatest for those systems which are seen edge on from the Earth; this maximises the radial velocities of the stars. The line separation will also be greater if the stars have relatively high orbital velocities; these can be worked out using Kepler's third law of planetary motion (it works for stars too). Kepler's third law is the 'complicated one' that relates the cube of the radius of the orbit to the square of the orbital period. Provided we stick to giving the masses of the stars 'M_1 and M_2' in solar masses and their separation 'a' in astronomical units (remember one astronomical unit or 1 AU is the mean distance from the Earth to the Sun and is equal to 1.5×10^8 km), then Kepler's third law gives in a very simple form the orbital period squared for the binary in sidereal years (one sidereal year equals 365.26 days);

$$P^2 = \frac{a^3}{M_1 + M_2} \qquad (4.5)$$

Best to do a simple calculation to show that it really is quite straightforward. Let's take two stars each of mass equal to that of the Sun which are orbiting each other at a distance of 1 AU. Plugging these value in gives P^2 equal to 1^3 divided by $1 + 1$ which equals $^1\!/_2$, so P is equal to $1/\sqrt{2}$, i.e. 0.707 sidereal years. The circumference of the orbit is $\pi \times$ the diameter; i.e. $\pi \times 1.5 \times 10^8$ km which equals 4.7×10^8 km. The stars have to cover this distance in 0.707 sidereal years; 0.707 sidereal years is equal to 2.2×10^7 s; so their orbital velocity has to be 4.7×10^8 divided by 2.2×10^7 which is equal to about 21 km/s. Now let's suppose that we 'see' the orbital plane of this binary edge on so the maximum radial velocity of the stars will also be 21 km/s; using the simple Doppler shift formula (4.2) we would observe a maximum shift for each of the stars of about 0.46 Å for the Hα line. So the lines' maximum separation would amount to a little under 1 Å. A smaller separation between the binary components and/or greater masses will result in a shorter orbital period and higher orbital and radial velocities.

Very often the masses of the two stars are unequal; this shifts the centre of mass of the system towards the more massive star. The two stars orbit the centre of mass with the same orbital period but now the less massive star is further away from the centre of mass and so has further to go; it therefore moves faster. As a result, the two stars have different orbital and hence radial velocities; what we observe is the two spectral lines oscillating about a central wavelength but with different amplitudes. A further level of complexity occurs when the stars move on elliptical orbits, again about the common centre of mass of the system. In this case the stars' orbital and hence radial velocities vary constantly and so the rate of separation of the lines varies too.

Summary

- Relative radial motion between a wave source and the observer causes a wavelength shift—the Doppler effect.
- For increasing separation between source and observer, the radial velocity is treated as positive and results in a red shift.
- For decreasing separation, the radial velocity is negative and a blue shift is the result.
- Motion across the observer's line of sight produces no wavelength shift.
- For radial velocities up to about 5000 km/s, the difference in wavelength shift when using the relativistic Doppler formula, as opposed to using the non-relativistic formula, amounts to less than 1 Å and so it's probably okay to use the simpler formula.
- The space motion of any object can split into two components; the tangential motion and the radial motion.

When Is a Spectral Line Not a Spectral Line?

Electron transitions enable atoms to absorb or emit light. From what we have learned so far, the absorbed or emitted light is of a very specific wavelength which itself depends on the energy difference between the energy levels involved in the transition. This leads to the conclusion that absorption lines and emission lines also have very specific wavelengths—the Hα line is at λ6563 after all. However, you only have to look at any astronomical spectrum to realise that spectral lines have a definite width, i.e. they spread out across a range of wavelengths. What's more by careful examination it becomes clear that some lines, if they are absorption lines, are darker than others and some may even appear to divide up into several dark component lines connected by less dark regions. A spectral line clearly isn't just a simple infinitely narrow line; the answer to the question posed in the title of this chapter is— 'when it is a *line profile*'. All spectral lines have a line profile; in older books it was also sometimes referred to as the line contour.

In this chapter we're going to dive into the wonderful world of physical processes which cause spectral lines to broaden. The reason for doing this is that these processes take place in stars and so careful study of broadened spectral lines can tell us many things about the stars themselves.

Line Profiles

In the old days, an astronomical spectrum was in fact a photograph of the spectrum itself recorded on a photographic plate. So it was often presented as a band of light (the continuum) running from violet to red, crossed by dark absorption lines or bright emission lines. Nowadays a spectrum is presented as a graph or a plot of intensity of radiation against wavelength. For most spectra the intensity is measured in arbitrary units which are probably some function of the CCD photon count; we then have what is known as an *uncallibrated spectrum*. Depending on the research being undertaken, professional astronomers will sometimes go to a great deal of trouble to secure spectra which are *calibrated*. This means that the plot is showing the actual intensity of radiation being received form the star and this is measured in watts per square metre per angstrom. Amateur spectra like most professional spectra are most likely to be uncallibrated spectra but they are still of course of enormous value because the topography of the spectrum plot doesn't depend on the units of intensity and it is the topography or shape of the spectrum which reveals so much about the physical processes which are taking place.

In a spectrum plot, absorption lines now appear as dips in the continuum; darker lines will be deeper dips and emission lines will appear as spikes superimposed on top of the continuum; the brighter the line the taller the spike. If we 'zoom in' on say one of the absorption lines, we will see that the line actually spreads out over a range of wavelengths and that instead of a 'line' we have perhaps a curved 'v' shaped dip in the continuum or maybe some other shape of curve. The actual shape of the curve is called the *line profile*. All spectral lines have such a line profile and there are several basic physical processes which produce it. Usually two or more of these processes combine together to give a spectral line its final profile, and in this chapter we'll be looking at the effects of these processes. There are also other processes which can produce some weird and wonderful line profile shapes in astronomical spectra and we'll look at some of these later in the book.

Equivalent Width

Some lines in a spectrum are broader and/or deeper (darker) than others and there is a straightforward way to quantify this. Fig. 5.1 shows a stylised absorption line as a rounded 'v' shaped dip in the continuum and in fact most line profiles will have a generally rounded shape with edges which gradually flatten out as they merge into the

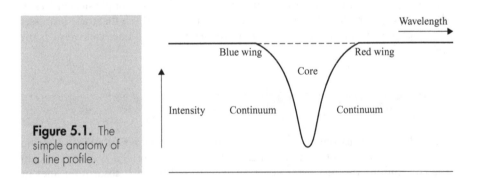

Figure 5.1. The simple anatomy of a line profile.

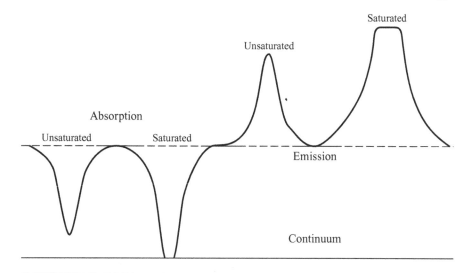

Figure 5.2. Unsaturated and saturated line profiles. The core of a saturated absorption line completely removes a narrow slice of the continuum. The top of a saturated emission line is flattened because over this narrow wavelength range, the hot gas is emitting the maximum amount of radiation possible, i.e. it is emitting as a black body.

neighbouring continuum. The edges or outer parts of a line profile are referred to as the *wings* while the central part of the profile is called the *core*.

Depending on how broad the line profile is and how deeply it penetrates into the continuum, there will be a smaller or greater area contained within the line profile. An absorption line which dips right down to the bottom of the continuum (such a line on an old style photograph spectrum would appear completely black) is said to be *saturated*. A saturated emission line profile is one which either goes right off the top of your plot or strictly speaking have a flattened top rather than a rounded one as shown in Fig. 5.2. This means that the actual source of that emission line is emitting as much radiation as is possible under the local conditions or put another way, just for that narrow range of wavelengths the source is emitting like a perfect black body.

Fig. 5.3 shows both a stylised absorption line and an emission line which enclose equal areas. Also shown is a rectangular 'slice' removed from the continuum whose area equals that within the two line profiles. The width of this rectangle in angstroms defines the *equivalent width* of the lines. This is a term which you will come across a great deal in the literature.

If you want to work out an equivalent width for yourself, one fairly straightforward way to determine the area within a line profile is to plot the spectrum on graph paper and count the squares. Alternatively, a computer mathematics package might be able to 'fit' or match a curve to the line profile and determine the area within it. It's also necessary to determine or at least estimate the height of the neighbouring continuum in the arbitrary units of intensity (determine the continuum height on either side of

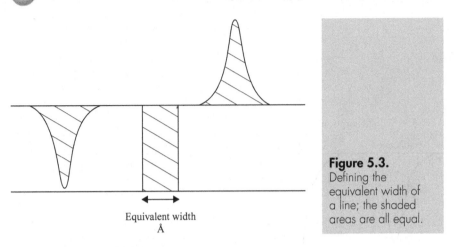

Figure 5.3.
Defining the equivalent width of a line; the shaded areas are all equal.

Equivalent width
Å

the line and take an average; the continuum may well be sloping across the width of the line). Now draw two parallel lines from the top to the bottom of the continuum such that the area contained within the lines (i.e. the rectangle) is equal to the area of the line profile. The height of the continuum will fix the width in angstroms of this artificial rectangular line profile and this equals the equivalent width of the real line profile. Finally, in case you have difficulty in determining where the line profile wings end and blend into the continuum, I'll be describing a method for doing this later in the chapter.

Populations of Atoms

A real astronomical spectrum is usually produced by some sort of gas; this might be the hot gas in the photosphere of a star or gas in a planetary nebula which is irradiated by ultraviolet radiation from the central star. Stellar photospheres and planetary nebulae are big, which means they consist of vast numbers or in other words, a vast population of atoms. Many of these atoms will be hydrogen atoms simply because the Universe contains a heck of a lot of this stuff; there will probably be plenty of helium atoms too as well as atoms of other chemical elements like carbon, oxygen and nitrogen. You could be forgiven for thinking though that within a given species of atom—say all the hydrogen atoms within a planetary nebula, each atom is pretty much the same as any other. They may be in different states of excitation of course with some having their electron in the ground state, some in the $n = 2$ level and so on. This could give rise to different spectral series; downward transitions to the $n = 1$ level will produce emission lines of the Lyman series in the ultraviolet, whereas transitions to the $n = 2$ level will give us the Balmer lines which of course include the Hα line. Surely though, all atoms which produce the Hα line are the same, aren't they? Well no! For one thing these atoms are all moving around in different directions and with varying speeds and this will affect the profile of the Hα line which is emitted by these atoms, as we'll see later. There are other things too which make atoms in such a vast population different from each other. We'll look at these things in turn and see how they affect the line profile.

Shivering Energy Levels

Let's do an imaginary experiment; a thought experiment or a gedanken experiment as physicists call it. Take a motionless hydrogen atom with its electron in the $n = 3$ level; the electron drops to the $n = 2$ level and the radiation comes out at $\lambda 6563$. Now take a seemingly identical hydrogen atom and let the same thing happen; again the radiation comes out at $\lambda 6563$—or does it? Well in fact, if we were careful in our measurements we'd find that the radiation doesn't come out at exactly $\lambda 6563$—it comes out at a wavelength slightly greater or less than this value. Yet another atom would give yet another slightly different result. If we repeated our thought experiment many times with individual atoms and combined our results, we'd find that there was a spread of wavelengths centred on $\lambda 6563$. All of these atoms are undergoing the same transition and therefore are producing Hα line radiation, so what's going on?

At the risk of it sounding like much of what was said in Chapter 3 was a 'con job'; the energy which an electron in a given energy level has is not a precisely defined quantity. This comes from another of the great principles of quantum mechanics called the *Heisenberg uncertainty principle*. In the situation here it says that the time an electron spends in an energy level and the energy it has when it's in that level are connected. An energy level which is very stable (a metastable level would be a good example) means that an electron will spend a long time there and in consequence the uncertainty in the electron's energy is very small. By contrast a fairly unstable level, i.e. one in which the electron would be expected to spend very little time (for example a higher level to which the electron has been excited), results in a greater uncertainty in the value of the electron's energy. The overall effect is that energy levels in an atom have an inbuilt 'fuzziness' and the less inherently stable a level is, the fuzzier it is as depicted in Fig. 5.4.

Imagine a vast population of hydrogen atoms and also imagine for the moment that they are not moving around but simply 'sitting there'. In this thought experiment our atoms all have their electrons in the $n = 3$ energy level; the electrons then drop to the $n = 2$ level and in doing so emit radiation which we use to produce a spectrum with our hypothetical spectroscope. All the radiation comes out close to $\lambda 6563$ and so our entire spectrum consists basically of just a Hα line. The radiation from some of our population of atoms may well come out at bang on $\lambda 6563$ but much of it will be emitted at wavelengths either slightly greater than or slightly less than the central value. The result is that our Hα line won't be a line at all but a broadened Hα line profile. It will look something like that shown in Fig. 5.5, i.e. a fairly narrow core with shallow but relatively broad wings. This form of spectral line broadening is an inherent property of all atoms and as described above comes from a fundamental principle of quantum mechanics—the Heisenberg uncertainty principle. It is called *natural line broadening*; it is also often called *natural line damping* and sometimes *radiation damping*. So if our population of atoms were doing nothing else, the spectrum produced by it would contain line profiles of finite width rather than infinitely thin lines.

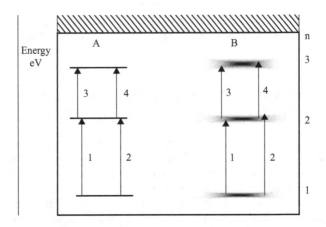

Figure 5.4. The effect of 'fuzzy' energy levels on electron transitions. In case 'A' the energy levels are sharply defined and so transitions 1 and 2 correspond to the same energy and hence wavelength, as do transitions 3 and 4. In case 'B' the uncertainty in the energy levels causes variation in the energies of transitions; furthermore the degree of 'fuzziness' increases for increasing levels of excitation, resulting in increased variation in wavelength for higher transitions.

It's easy to see how this process works for absorption. If our hydrogen atoms were bathed in a beam of light (the correct general term for a beam of any form of electromagnetic radiation is a *radiation field*), then some of this light would be absorbed by the atoms. Let's suppose that initially all the atoms have their electron in the $n = 2$ level so that light from the radiation field at $\lambda 6563$ will be absorbed to give us the centre of the Hα line. However, due to the slightly different energy values ascribed to the energy levels (the $n = 2$ and $n = 3$ levels in this particular case) within our population of atoms, light at wavelengths on either side of $\lambda 6563$ will be absorbed too and this will result in a naturally broadened Hα absorption line.

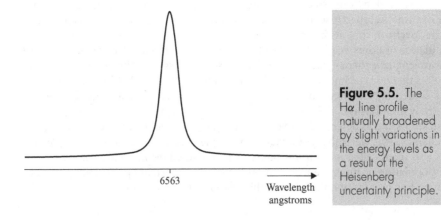

Figure 5.5. The Hα line profile naturally broadened by slight variations in the energy levels as a result of the Heisenberg uncertainty principle.

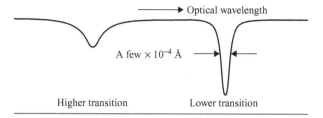

Figure 5.6. The shapes of naturally broadened line profiles; they are narrow but with relatively extensive wings for transitions to lower energy levels (e.g., the Hα line) but become more smeared out for transitions to higher levels (Hβ, Hγ, etc.). Even so the overall effect is very slight; amounting to no more than a few ten thousandths of an angstrom at optical wavelengths.

The higher the excitation of an energy level, the shorter the length of time an electron is likely to spend in it. This means a greater uncertainty in the electron's energy and in consequence a broader line for transitions involving higher levels in a transition series as illustrated in Fig 5.6. However, for a given number of atoms a broader line also means a shallower absorption line or a less intense emission line. Having said all this it turns out that at optical wavelengths the uncertainties in energies correspond to wavelength spreads of the order of a few ten thousandths of an angstrom; so natural line broadening isn't something which is likely to cause the amateur spectroscopist any problems.

Enter the Doppler Effect

As described in the previous chapter the Doppler effect is one of those bits of physics that crops up time and again in astronomy, which is another way of saying that it has enormous value for astronomers. In the context of astronomical spectroscopy, an atom emits a photon of light as a result of an electron transition (let's stick with the $n = 3$ to $n = 2$ transition for hydrogen) and the emitted photon is observed to have a wavelength of λ6563. Now let the atom move so that a component of its motion is directed either towards or away from us; remember this component of the atom's motion is called the *radial motion* and the associated velocity is called the *radial velocity*. Then as is well known, if the radial velocity is directed towards us, the emitted photon will be observed to have a wavelength shorter than λ6563, i.e. the photon has been blue shifted. Conversely, if the radial velocity is directed away from us the emitted photon is seen to be red shifted. The greater the radial velocity is, the greater is the resulting blue or red shift. As we saw in Chapter 4, provided the radial velocity isn't too high, the equation connecting the shift in observed wavelength; Δλ with radial velocity v, is simply given by

$$\Delta\lambda/\lambda = v/c \qquad (5.1)$$

Here λ is the unshifted wavelength (6563 Å in this case) and c is the speed of light.

When Things Get Hot

The individuals in the vast population of atoms making up the gas in say the atmosphere of a star or a planetary nebula are in constant motion. This is simply because the gas has a temperature above absolute zero and in fact the higher the temperature of the gas, the greater the average speed of the gas atoms. Notice the phrase here 'average speed'; like the crowd of runners in the New York marathon, some have higher speeds than others. The range of speeds for a population of atoms results in an equivalent range of radial velocities as seen by us here on the Earth. There will be a range of radial velocities directed both towards us and away from us. There will also of course be many atoms which are moving across our line of sight and which thus have zero radial velocity. Let's say we have a population of hydrogen atoms and let's also suppose that we have been able to somehow eliminate natural line broadening. Hα photons emitted by our population will show a range of red and blue shifts on either side of λ6563 and this will again result in a broadened line profile rather than an infinitely thin line.

The line profile itself will not surprisingly reflect the range and distribution of velocities of the atoms and in fact the profile turns out to have a shape which is very familiar to statisticians. It is that of what is known as a *normal distribution* or sometimes a *Gaussian distribution* (named after the great nineteenth-century German mathematician—Carl Friedrich Gauss). It is a bell-shaped profile like those shown in Fig. 5.7, which I've computed for some typical stellar temperatures. It has a wider core but less well-developed wings than the natural line broadening profile. The width of the core itself is determined by the average velocity of the gas atoms which in turn is determined by the temperature of the gas. This form of line broadening which

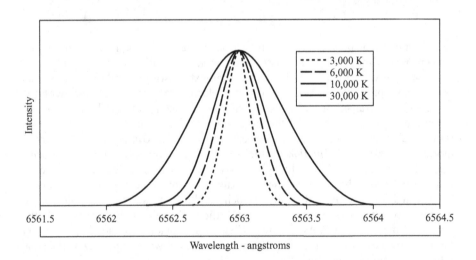

Figure 5.7. Thermally broadened Hα line profiles for a range of temperatures; the characteristic 'bell-shaped' profile is well known to statisticians as a 'normal distribution' or sometimes a 'Gaussian distribution' after Carl Friedrich Gauss.

is caused by the thermal motion of the gas atoms is called not surprisingly *thermal broadening* and sometimes also *Doppler broadening*.

For the case of absorption lines, we have to imagine ourselves riding on a hydrogen atom. We are bathed in the light of the radiation field which of course includes photons at λ6563. Because of our motion however our hydrogen atom 'sees' these photons at a different wavelength, Doppler shifted by our radial velocity relative to the incoming light beam. So our electron in the $n = 2$ level ignores them and they pass us by. However that same Doppler shift can make a photon of some other wavelength appear as λ6563 and our atom absorbs the photon. Back on the Earth it appears that this particular atom has absorbed a photon in the vicinity of, but not exactly at λ6563 and indeed other atoms will be perceived to have absorbed photons covering a range of wavelengths centred on λ6563 to produce a thermally broadened absorption line.

How to Determine the Temperature of the Gas

In most cases the width of the core of a Doppler broadened line profile will be significantly greater than that due to natural line broadening; as we have seen, natural broadening amounts to perhaps a few ten thousandths of an angstrom. So for the most part we can ignore it and then it turns out to be very easy to work out the temperature of the gas which is responsible for a thermally broadened spectral line. A population of gas atoms will have a range of velocities which depend on the temperature of the gas. Within this range there will be one velocity which the gas atoms are more likely to have than any other and this is often called the *Doppler velocity* v_D. The Doppler velocity is related to the temperature of the gas by (once again) a very simple formula.

$$v_D = \sqrt{\frac{2kT}{m}} \tag{5.2}$$

Here, k is one of the great fundamental constants of physics called *Boltzmann's constant* named in honour of the Austrian physicist Ludwig Edward Boltzmann. It is *the* constant which connects the temperature of a gas to the speed of its atoms or molecules and its value is equal to 1.381×10^{-23} J/K. The quantity m in Eq. (5.2) is the mass of the individual gas atoms we are dealing with; for example, the mass of the hydrogen atom is 1.674×10^{-27} kilograms (kg) and T of course is the temperature (in Kelvin) which we want to work out.

A radial velocity of v_D corresponds to a wavelength shift which we can call $\Delta\lambda_D$ and if we plug this and v_D into Eq. (5.1) we have

$$\Delta\lambda_D = \lambda_0 \times v_D/c \tag{5.3}$$

Using the value for v_D given in Eq. (5.2), a bit of rearranging will give us the temperature T in terms of $\Delta\lambda_D$

$$T = \frac{m \times c^2}{2 \times k \times \lambda_0^2} \times \Delta\lambda_D^2 \tag{5.4}$$

Let's put the mass of the hydrogen atom in for m and the values for c and k in order to start turning this into a simple 'plug in' formula; we get

$$T = 5.455 \times 10^{12} \times \frac{\Delta\lambda_D^2}{\lambda_0^2} \tag{5.5}$$

The quantity $\Delta\lambda_D$ turns out to be related in a very simple way to a quantity which can be measured directly from a spectrum; this is the total width in angstroms of a line profile at half of its maximum height or depth. This is called *the full width half-maximum* (FWHM). $\Delta\lambda_D$ is equal to FWHM/1.665; so $\Delta\lambda_D^2$ is then equal to (FWHM)^2/2.772. This turns Eq. (5.5) into

$$T = \frac{1.968 \times 10^{12} \times (\text{FWHM})^2}{\lambda_0^2} \tag{5.6}$$

Let's take our old friend the Hα line; so λ_0^2 will equal 6563 $\times$ 6563. Suppose the FWHM for the line is 1 Å, then plugging these values in we get $T = (1.968 \times 10^{12} \times 1^2)/(6563 \times 6563)$ which is approximately equal to 45,000 K. Don't worry too much about the above details unless you need to use the mass for atoms other than hydrogen in Eq. (5.4) (all other atoms being more massive than hydrogen, will give narrower line profiles for a given temperature), otherwise just use Eq. (5.6).

Having said all this, the chances are that you'll already know the temperature of the star simply from its spectral type. However, you may well find that the FWHM of a spectral line is greater than it should be based on this temperature and this is most likely due to turbulence in the star's atmosphere.

Turbulence

In the outer layers of stars (including the Sun) heat from inside the star is carried to the surface primarily by *convection*; that is to say, 'globs' of gas pick up heat from the interior and rise to the surface. The heat is radiated away; the globs of gas cool and sink back down into the interior. In many cases, particularly in the atmospheres of red giant stars, this convection process is not a smooth one; the atmospheric gases suffer a great deal of turbulence. Turbulence is caused by larger scale motions of the gas atoms which give rise to swirls and eddies; the same kind of thing happens in the Earth's atmosphere and in flowing streams and rivers. The overall result is random motion just like thermal motion but on a larger scale and what's more the effect on spectral lines is exactly the same as Doppler broadening. Thus, turbulence will simply increase the width, i.e. the FWHM of a thermally broadened line. Just as we used the Doppler velocity v_D to quantify the thermal motion of the gas atoms we can do the same for the effect of turbulence and simply call it the *turbulence velocity* 'v_T'. The turbulence velocity is the most likely velocity that the gas atoms will have due turbulence phenomena in the star's atmosphere and this simply adds to the Doppler velocity. So in Eq. (5.3), v_D is replaced by $v_D + v_T$ and Eq. (5.3) turns into

$$\Delta\lambda_{D+T} = \lambda_0 \times (v_D + v_T)/c \tag{5.7}$$

As before $\Delta\lambda_{D+T} = $ FWHM/1.665. If you know the temperature of the star (e.g. from its spectral class), then Eq. (5.2) will give you the Doppler velocity v_D and a bit of rearranging of Eq. (5.7) will enable you to calculate the velocity due to turbulent motion in the star's atmosphere.

Piling the Pressure On

One of the things which is clearly going to affect the equivalent width of a spectral line is simply the number of available atoms to either absorb or emit radiation of a given wavelength. If the gas is of fairly low density; i.e. relatively few atoms per cubic metre, then 'collisions' between atoms will be relatively few and far between. The word 'collision' though widely used is not a good one; a better term would be 'close encounter'. With continued increase of the gas density, close encounters between atoms become more frequent. An atom is a tiny bundle of electricity and when two atoms get close, their electric fields affect, or as the professionals say, 'perturb' each other. This perturbing of the atoms' electric fields disturbs the energy levels in the atoms and results in transitions between a given pair of levels taking place at wavelengths which are shifted from that which would normally be expected. In the usual vast population of atoms this results in a broadened line profile. This form of line broadening is known as *collision or pressure broadening.*

It turns out to be a very complex process and ironically hydrogen which of course in many respects is spectroscopically the simplest of atoms, is the most difficult to analyse for collision broadening. This is largely due to the degeneracy of the *l* sublevels. Generally speaking though the effect of collision broadening on the line profile is similar to that of natural line broadening but on a much bigger scale, i.e. most of the broadening takes place in the profile wings. Collision broadening is as might be expected, important in stars with dense atmospheres, notably white dwarfs, it is also important as we shall see later in the study of what's called the 'curve of growth'.

Convolutions

On paper a spectrum looks just like a graph which plots intensity of radiation against wavelength and it is tempting to think that you could determine how much radiation is being emitted at any given wavelength by simply reading values directly off the plot. In reality though your spectrum consists of a series of 'wavelength bins', ostensibly of equal width running along the wavelength axis; each bin is centred on a specific wavelength. All incoming photons whose wavelengths fall within the limits spanned by a given bin will go into that bin. When the spectrum is complete, all of these photons will add to produce the total contribution from the bin and this in turn will produce one single point on the final spectrum plot as depicted in Fig. 5.8. If the wavelength bins are narrow the final spectrum will be of high resolution and will show a lot of detail; on the other hand if the bins are wide, more information will be lost because all the photons which fall within this wavelength range are simply being added together to make one point.

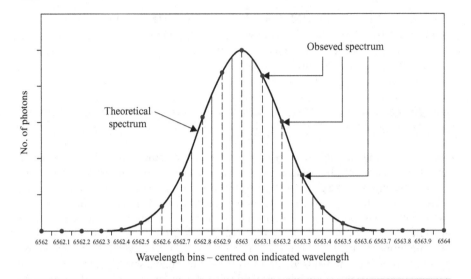

Wavelength bins – centred on indicated wavelength

Figure 5.8. A real spectrum consists of a series of 'wavelength bins' centred on regularly spaced wavelengths; each point on the spectrum corresponds to the number of photons whose wavelengths fall within the range of the wavelength bin. Narrower wavelength bins (i.e. a higher resolution spectrum) result in more closely spaced points and an observed spectrum which approaches the theoretical one.

The instruments which you use to produce a spectrum; i.e. telescope, spectroscope, CCD camera will combine to produce wavelength bins which will never be narrower than some minimum width. The result is that there will always be some blurring and therefore loss of fine detail in the spectrum. This is irrespective of other line broadening mechanisms and of course atmospheric seeing; it is called the *instrumental profile*.

Now take a spectral line with significant thermal broadening and a high-resolution spectrum; clearly the line profile will be spread over several wavelength bins. Let's think about what's happening in a typical wavelength bin. It contains those photons whose wavelength falls within the range of the bin. These come from atoms whose radial velocities fall within a given range as seen by the observer on the Earth. So their combined contribution to the line profile is just one single point on the plot. But hang on a minute; our subset population still amounts to a huge number of atoms and this subset of course is subject to natural line broadening. So while the range of radial velocities within the subset will put their contribution into one single wavelength bin in the spectrum, the added effect of natural line broadening within the subset enhances, i.e. increases the range of wavelengths which are either absorbed or emitted by this subset population. For many atoms, their contribution will still fall within our wavelength bin but those which would come close to the edge of the bin due just to their radial velocity can now be pushed over the edge into the next bin by the effect of natural line broadening. Conversely, contributions from the bins next door can be pushed over into our bin by the same effect. This is shown in Fig. 5.9.

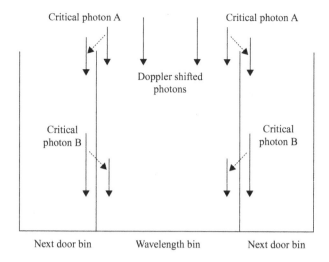

Figure 5.9. The total number of photons in a typical wavelength bin contributes to one single point on the spectrum. Photons, whose wavelength due just to Doppler broadening would put them close to the edge of the bin, can in fact be 'pushed over' into the next door bins by the effect of natural broadening (critical photons A), which can shift their wavelength just that bit more. In turn, the bin can receive photons from the next door bins by the same process as with the critical photons B depicted here.

Our thermally broadened line profile effectively has a series of natural line broadened profiles running through it. We say that the Doppler profile is *convolved* with a natural broadening profile. A thermal or Doppler broadened profile which is convolved with a natural broadening profile will have slightly extended wings (as mentioned above the width of the natural line broadening core is usually very much less than that of the Doppler core) and this profile a called the *Voigt* (pronounced 'foyked') *profile* after the German astronomer Hans-Heinrich Voigt.

How Broad Is a Line Profile?

If you take or are planning to take high-resolution spectra, it's very likely that you're interested in looking at line profiles in more detail. We've already seen how measuring the full width half-maximum for a line can give information on the temperature of a gas well as the velocity due to turbulent motion in a star's atmosphere. Provided you can estimate the level of the neighbouring continuum it's not too difficult to do this. However, a seemingly much more difficult problem is to determine exactly where the wings of a line profile end and blend into the continuum. This is particularly the case if the neighbouring continuum has lots of little spikes and dips in it which may be real spectral features or simply 'noise' due to the instrumentation. Here's a method for determining the wing limits of a line profile which seems to give good results

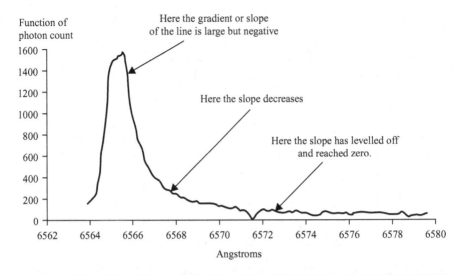

Figure 5.10. A typical line profile from core to red wing; the gradient or slope of the line becomes less negative as we approach the wing and eventually reaches zero as it merges into the continuum.

(it certainly satisfied the examiners of my PhD thesis); to use it you really need a computer spreadsheet package such as Microsoft Excel.

The raw data for a spectrum is just a series of pairs of numbers; one of these numbers will be a wavelength and the other a number whose value is related in some way to the photon count produced by the equipment. In fact, for want of a better name we can simply call this number the photon count; so the spectrum is just a plot of photon count vs. wavelength. We'll take the case here of an emission line but the method works equally well for absorption lines. We'll also assume for the moment that the continuum across the width of the line is 'flat'; this means that there is no overall continuum slope across the line irrespective of all the little spikes and dips. We shall also cut the line in two so that we can deal with the red and blue wings separately; so let's plot the red wing of our line profile as shown in Fig. 5.10.

As we move away from the top or peak of the line, we plunge down the slope of the line core. This part of the line has a significant negative gradient which is another way of saying that the photon count is dropping fast as we increase the wavelength. As we approach the wing this gradient levels off and becomes less negative. Eventually the line will reach the continuum and the gradient will approach zero. So the key to finding the wing limit is to find where the gradient of the line becomes zero. Clearly we need to be able to measure the gradient or slope of the line as we move along it from peak to wing and this is most easily done with a computer spreadsheet.

Have your data in two columns as shown in Fig 5.11a; then at the head of a new column use the spreadsheet 'slope' function (all spreadsheets should have a version of this function) to determine the slope or gradient of the line profile over a number of wavelength bins; here I've used 10 bins.

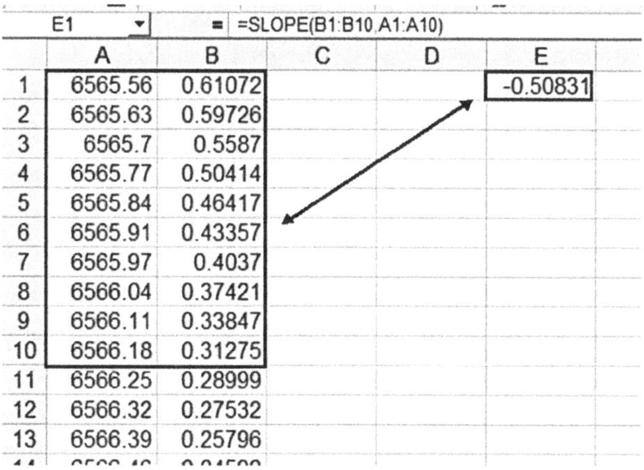

Figure 5.11.a.
Use the 'slope' function on your spreadsheet to calculate the gradient (shown here in cell E1) of the line profile over a small range of wavelength bins.

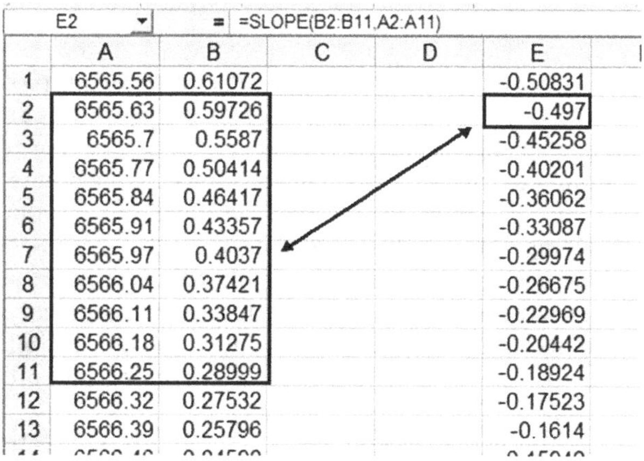

	A	B	C	D	E
E1 ▾	=	=SLOPE(B1:B10,A1:A10)			
1	6565.56	0.61072			-0.50831
2	6565.63	0.59726			
3	6565.7	0.5587			
4	6565.77	0.50414			
5	6565.84	0.46417			
6	6565.91	0.43357			
7	6565.97	0.4037			
8	6566.04	0.37421			
9	6566.11	0.33847			
10	6566.18	0.31275			
11	6566.25	0.28999			
12	6566.32	0.27532			
13	6566.39	0.25796			

Figure 5.11.b.
Here I've clicked on cell E1 and dragged the mouse all the way down the data set. I've then used the 'fill down' operation to determine the gradients of successively shifted line profile segments. The gradient of the boxed segment is given in cell E2 and so on.

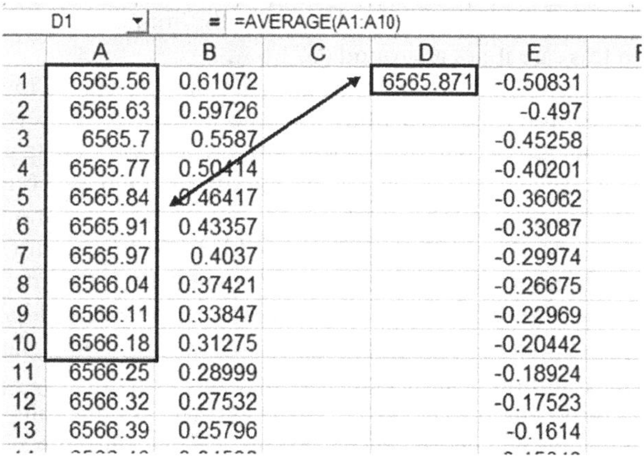

	A	B	C	D	E	
E2 ▾	=	=SLOPE(B2:B11,A2:A11)				
1	6565.56	0.61072			-0.50831	
2	6565.63	0.59726			-0.497	
3	6565.7	0.5587			-0.45258	
4	6565.77	0.50414			-0.40201	
5	6565.84	0.46417			-0.36062	
6	6565.91	0.43357			-0.33087	
7	6565.97	0.4037			-0.29974	
8	6566.04	0.37421			-0.26675	
9	6566.11	0.33847			-0.22969	
10	6566.18	0.31275			-0.20442	
11	6566.25	0.28999			-0.18924	
12	6566.32	0.27532			-0.17523	
13	6566.39	0.25796			-0.1614	

Figure 5.11.c.
Now use the 'average' or 'mean' function to determine the central wavelength of your first line profile segment; entered here in cell D1.

	A	B	C	D	E	F
D1 ▾	=	=AVERAGE(A1:A10)				
1	6565.56	0.61072		6565.871	-0.50831	
2	6565.63	0.59726			-0.497	
3	6565.7	0.5587			-0.45258	
4	6565.77	0.50414			-0.40201	
5	6565.84	0.46417			-0.36062	
6	6565.91	0.43357			-0.33087	
7	6565.97	0.4037			-0.29974	
8	6566.04	0.37421			-0.26675	
9	6566.11	0.33847			-0.22969	
10	6566.18	0.31275			-0.20442	
11	6566.25	0.28999			-0.18924	
12	6566.32	0.27532			-0.17523	
13	6566.39	0.25796			-0.1614	

	A	B	C	D	E
D2	▼	= =AVERAGE(A2:A11)			
1	6565.56	0.61072		6565.871	-0.50831
2	6565.63	0.59726		6565.94	-0.497
3	6565.7	0.5587		6566.009	-0.45258
4	6565.77	0.50414		6566.078	-0.40201
5	6565.84	0.46417		6566.147	-0.36062
6	6565.91	0.43357		6566.215	-0.33087
7	6565.97	0.4037		6566.283	-0.29974
8	6566.04	0.37421		6566.352	-0.26675
9	6566.11	0.33847		6566.421	-0.22969
10	6566.18	0.31275		6566.49	-0.20442
11	6566.25	0.28999		6566.559	-0.18924
12	6566.32	0.27532		6566.628	-0.17523
13	6566.39	0.25796		6566.696	-0.1614
14	6566.46	0.24592		6566.764	-0.15042

Figure 5.11.d.
Again click, drag
and 'fill down' to
determine the
remaining central
wavelength values.

The next thing to do is to move our segment of line profile along by one wavelength bin and calculate the slope again. We then repeat this process through the whole of the profile and this entire operation is done with phenomenal ease by clicking on your first calculated slope value and dragging the mouse down to the bottom of the data set. Now simply use the 'fill down' operation on your spreadsheet to calculate all remaining slope values as shown in Fig 5.11b.

Now use the 'average' or 'mean' function on your spreadsheet to determine the central wavelength of your first line profile segment as show in Fig 5.11c and again click and drag the mouse down, and use the 'fill down' operation to determine the central wavelengths of all remaining line profile segments as shown in Fig 5.11d.

You now have two new columns of numbers; the central wavelengths of successive line profile segments together with the gradients of these segments. If you run your eye down the gradient column, you'll see that an initially large negative value gets smaller and at some point it 'flips over' and becomes positive as shown in Fig 5.12. This is the point you're looking for—the limit of the line profile wing. As you can see in this case it lies at around 6571.6 Å.

Finally, Fig 5.13 shows what a plot of the slope of the line profile looks like and confirms our estimate of the wing limit.

You may need to experiment with the number of data points which you use in each line profile segment but it's very important that these points cover as small an interval on the line profile as possible. This will enable you to monitor the gradient of the line more closely; theoretically you could use just two consecutive points, but this may result in too much fluctuation in the gradient of the line. If there is an obvious slope in the continuum across the line profile then use the 'slope' function to determine the gradient of the continuum on either side of the line. Take an average to determine the continuum slope across the width of the line and in your plot of line gradient vs. wavelength, read off where the plot reaches this value. This may be a negative or positive value depending on the slope of the continuum.

	A	B	C	D	E	F
77	6570.78	4.74E-02		6571.092	-0.03097	
78	6570.85	4.38E-02		6571.161	-0.03847	
79	6570.92	3.63E-02		6571.23	-0.05018	
80	6570.99	3.01E-02		6571.299	-0.05964	
81	6571.06	3.07E-02		6571.368	-0.05974	
82	6571.13	3.70E-02		6571.436	-0.04827	
83	6571.2	3.84E-02		6571.504	-0.02725	
84	6571.26	3.44E-02		6571.572	-0.0017	
85	6571.33	2.62E-02		6571.641	0.024901	
86	6571.4	1.96E-02		6571.71	0.049023	
87	6571.47	9.12E-03		6571.779	0.06527	
88	6571.54	-1.80E-04		6571.848	0.066663	
89	6571.61	1.93E-03		6571.917	0.04948	
90	6571.68	1.33E-02		6571.985	0.02576	
91	6571.74	2.42E-02		6572.053	0.010651	

Figure 5.12. Looking down columns D and E, we see that between λ6571.572 and λ6571.641 the gradient 'flips over'; i.e. changes from negative to positive. In doing so it passes through zero at around λ6571.6 and this marks the limit of the line profile wing.

This method works pretty well for spectra which are reasonably 'clean' and where a line is strong compared to the neighbouring continuum. The method is harder to use when the line is relatively weak compared to a 'noisy' continuum but at least it can give an estimate of the overall width of the line.

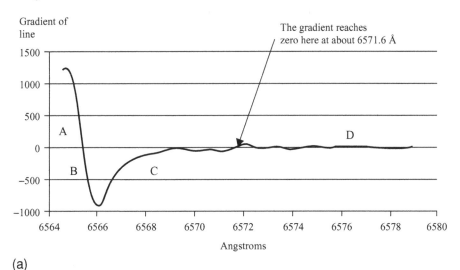

(a)

Figure 5.13. (**a**) A plot of the line profile's gradient vs. wavelength. The zones marked A, B, C and D correspond to similarly marked zones on the line profile plot in (**b**).

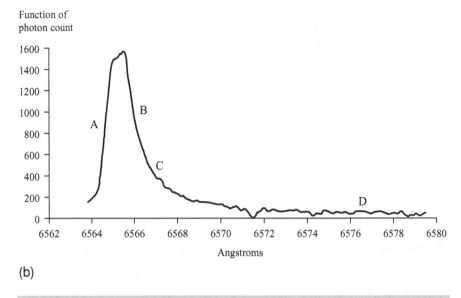

(b)

Figure 5.13. (*Continued*)

Summary

- Spectral lines have a characteristic 'shape' which is called the line profile.
- Line profiles are broadened primarily by four basic processes.
- Natural line broadening and pressure broadening tend to affect the line profile wings.
- Thermal and turbulence broadening tend to broaden the line profile core.

Stellar Spectra and That Famous Mnemonic

Sooner or later, every amateur astronomer gets to know about the spectral classes of stars and that the letters which represent the spectral classes; OBAFGKM, can be remembered with that famous mnemonic; 'Oh Be A Fine Girl Kiss Me'. In fact every self-respecting astronomer should ask the question; 'Why such an odd jumble of letters? Why not ABCD etc.?' I myself can't remember asking such questions—well it *was* a long time ago!

In this chapter we'll learn something about the history of why the spectral sequence got to be this way but first we need to have a look at why stellar spectra are the way they are. This, as you might expect involves looking at the physical processes going on within the surface layers of stars.

Stellar Atmospheres

Take a look at any star on a dark clear night; it appears as a tiny yet intense point of light and you tell yourself that the light which you are seeing began its journey many years ago because that's how long it's taken that light to travel across the light years of space. The story of that light is in fact even more remarkable; the light you see is a stream of photons which pass through the lens of your eye to the retina. Since leaving the star, the photons have dodged interstellar dust grains, comets in the Oort cloud, Kuiper belt objects, planets, asteroids and molecules in the Earth's atmosphere. It's been a tough journey but compared to what happened before it left the star, pretty easy really.

These photons started their lives as gamma ray photons deep inside the star's core; they then had to make their way through the hotter deeper denser layers of the star, constantly being absorbed and re-emitted by intervening atoms. This they did in a random walk fashion which took them millions of years; this 'walk' takes its toll and by the time they reach the outer layers of the star, they have lost a lot of their energy which itself has been used to help heat up the vast body of the star itself. As they near the surface of the star they have become photons of visible light and this is where their journey gets really interesting.

Meanwhile, back in your backyard, you've set up your spectroscope in order to acquire a spectrum of the star which you see consists of a continuous band of colour crossed by dark lines. The band of colour which is called the *continuum* clearly results from those photons which have made it and left the star's surface. As for the dark lines—the *absorption lines*, well clearly something has happened to them. Immediately you recall the work of Kirchoff and Bunsen which we described back in Chapter 2; an absorption spectrum is produced when a continuous spectrum is made to shine through an intervening layer of cooler gas. What chemical elements the gas is made of will determine where the absorption lines are. In the case of our star, the continuum must come from the denser deeper hotter layers, while the absorption lines are caused by cooler thinner gas in the star's surface layers. This was the initial idea as to how absorption lines get formed in stellar spectra; the layer of cooler gas which does the absorbing was called the reversing layer. Nowadays ideas or models (astrophysicists use the laws of physics to construct model stellar atmospheres) are a bit more sophisticated than this simple model though in many ways the basic idea is still the same. We can also avoid getting involved in some serious mathematics by simply thinking about some of the things that could have happened to our lost photons within these outer layers of a star.

Continuous Absorption

If it were possible to send a probe carrying a spectroscope deep down inside the atmosphere of a star, we would find the spectrum produced by these deeper layers became increasingly like that of a perfect black body. It's when we approach the surface that 'imperfections' appear in this otherwise almost perfect spectrum. Absorption lines are of course one form of imperfection but there are some processes which can knock out whole chunks of the continuum.

Photoionisation

As before, let's stick with hydrogen to keep things as simple as possible. Hydrogen atoms with their electrons in the $n = 2$ energy level can absorb photons at selected wavelengths to give us the Balmer series of absorption lines in the visible part of the spectrum. Shorter wavelength photons will send the electrons to ever higher energy levels within the hydrogen atoms but a photon at λ3647 will send an electron from the $n = 2$ level out into the 'free world', i.e. it will ionise a hydrogen atom from this level. In fact, all photons with wavelengths shorter than λ3647 will ionise hydrogen atoms which are in the $n = 2$ excited state. Another way of stating this is to say that these hydrogen atoms are potential targets for photons coming up from the star's interior—and a direct hit means of course that the photon is destroyed!

A photon with just about the right kind of energy to just about ionise one of these atoms in a sense 'sees' an approaching atom as a big target. Photons of higher energy; i.e. shorter wavelength 'see' the atoms as ever diminishing targets so as the wavelength decreases, more and more of the photons get through. The result is that, particularly for hot stars which radiate plenty of light at the short wavelength end of the spectrum, the continuum rises gradually as we head towards this region but then drops suddenly as photons at just the critical wavelength get absorbed by the hydrogen atoms. This catastrophic drop in the continuum is called the *Balmer jump* and sometimes the *Balmer discontinuity*. The continuum then recovers and rises again as we move away from the critical wavelength and into the ultraviolet. The result is a saw-tooth notch cut out of the otherwise smooth continuum. At 3647 Å the Balmer jump is situated in the blue-violet part of the spectrum; hydrogen atoms in the $n = 3$ excited state can be ionised with photons at $\lambda 8212$ which is in the near infrared part of the spectrum. This is the Paschen jump which corresponds to the Paschen series of lines in the infrared. The piece of spectrum from the Paschen jump to the Balmer jump is often referred to as the *Paschen continuum* just as the continuum to the short wavelength side of the Balmer jump is referred to as the *Balmer continuum*.

Photoionisation of hydrogen is the main cause of continuous absorption in hotter stars. The higher temperatures in the outer layers of these stars causes most of the hydrogen atoms there to be in an excited state with the electrons in the $n = 2$ or

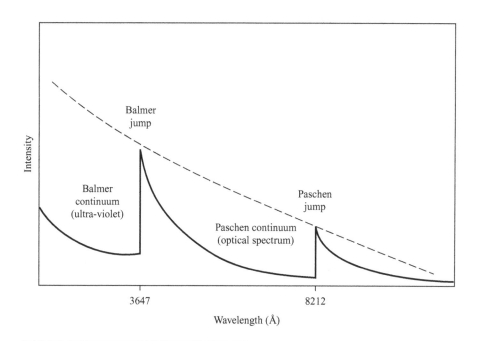

Figure 6.1. The effect of transition 'jumps' on a star's continuous spectrum; the dotted line approximates a black body radiation curve for the temperature of the star.

higher levels. For cooler stars most of the hydrogen atoms are in the ground state with the electrons in the $n = 1$ level. These atoms can only be ionised by ultraviolet photons so photons in the optical part of the spectrum are spared. Similar jumps can occur due to other elements but because of the overwhelming dominance of hydrogen in stellar atmospheres, the effects of these are usually quite small.

The Negative Hydrogen Ion

Hydrogen is simple; its atoms consist of a single positively charged proton together with a single negatively charged electron. However, hydrogen atoms can and do sometimes have two electrons; they are then called *negative hydrogen ions* and are written H^-. This might seem at first a little bizarre unless you are a chemist but it turns out that the electric field of the hydrogen nucleus is not totally screened off by the single electron and it's possible for the atom to capture a second electron. This electron is however only weakly bound and it doesn't take much energy to remove it; in fact only 0.754 eV which can be provided by photons with a wavelength shorter than about $\lambda16,450$. This wavelength is well into the infrared and so clearly all visible light photons will remove the extra electron (the term used here is *dissociation* rather than ionisation). It also means that H^-ions can only exist in the atmospheres of cooler stars because the higher temperatures in hotter stars will quickly dissociate any H^- ions which form.

Photoionisation of neutral hydrogen resulted in the removal of a saw-tooth shaped piece of continuum; with photodissociation of H^-, the effect is a bit different. Absorption of photons begins gradually at $\lambda16,450$ and gradually increases as we go to shorter wavelengths until the absorption peaks at about $\lambda8500$. Absorption then decreases gradually towards shorter wavelengths and the overall result is a 'saucer shaped' swathe removed from the continuum.

The dissociation of the H^- ion is clearly a form of bound–free transition; however, just as comets can orbit the Sun on open-ended parabolic and hyperbolic orbits, it's possible for an electron to orbit a neutral hydrogen atom in the same way. While the electron is doing this, it can absorb an incoming photon which sends it into a different hyperbolic orbit. The incoming photon has then been lost in a free–free transition. This form of free–free transition can also happen to electrons which are having close encounters with other kinds of atoms. The effect of these free–free transitions is difficult for professional astronomers to calculate, but they shouldn't concern amateur spectroscopists too much due to the fact that their greatest effect is in the infrared, because free–free electron transitions tend to involve exchanges of small amounts of energy.

Broadly speaking then, hotter stars suffer continuous absorption in their spectra as a result of photoionisation of neutral hydrogen in their atmospheres. In cooler stars (this includes the Sun) the presence of negative hydrogen ions is the main cause of continuous absorption.

Line Absorption

Now for the absorption lines; clearly in order for absorption lines due to some chemical element to be present in the spectrum of a star, the element has to be there in the first place. It would then seem natural to suppose that the more abundant a given element was, the stronger its absorption lines would be. Well not necessarily! Let's see why.

It's safe to say that hydrogen is abundant in the atmosphere of every star (though some stars are known to be hydrogen deficient) and its presence is revealed by the Balmer lines in the optical part of the spectrum. To make a Balmer line we need not just any hydrogen atom but a hydrogen atom with its electron in the $n = 2$ level. When hydrogen is cool the electrons stay in the $n = 1$ or ground level, so cool hydrogen cannot make Balmer lines even if there are vast amounts of the stuff present. Heat the hydrogen up and thermal excitation makes the electrons migrate up to the next level and by the time the gas reaches a temperature of 10,000 K, the $n = 2$ level has become maximally populated. Hydrogen at a temperature of 10,000 degrees then is the stuff for making Balmer lines and it is no coincidence that 10,000 degrees is the temperature in the surface layers of stars of spectral class A. Spectral class A stars have the strongest Balmer lines in the entire spectral sequence; cooler stars have fewer and fewer hydrogen atoms in the correct state to produce Balmer line absorption and so the lines become weaker. If the hydrogen heats up to higher temperatures, the electrons migrate to higher energy levels—the Paschen lines in the infrared will become stronger at the expense of the Balmer lines. Eventually the hydrogen will become largely ionised and all hydrogen absorption lines will fade as they do for the very hottest stars.

Fig. 6.2 shows how the $n = 2$ level in hydrogen becomes increasingly populated as the temperature rises. This plot was computed using two fairly involved equations; one is called Boltzmann's equation and this gives the number of atoms in the $n = 2$ level as a fraction of all the neutral atoms, in terms of the temperature. The other equation (even more complicated) is called Saha's equation and this gives the number of ionised atoms as a fraction of the total number of atoms, again in terms of the

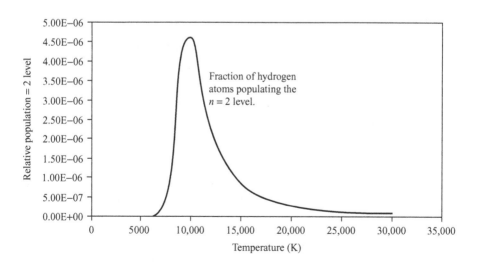

Figure 6.2. The fractional population of the $n = 2$ level for hydrogen as a function of temperature; the population of this level by electrons clearly peaks around 10,000 K giving rise to the strongest Balmer lines in the spectra of class A stars.

temperature. By combining Boltzmann's equation with Saha's equation it is possible to calculate the number of atoms in any given level as a fraction of the total number of atoms and this is what's plotted here. Notice how the graph rises quickly for increasing temperature and peaks at around 10,000 K; it then falls off more gradually for higher temperatures. Full details of both Boltzmann's equation and Saha's equation can be found in more advanced books.

All of this tells us two things; firstly that the earliest attempts at Harvard around the turn of the nineteenth to twentieth centuries, to produce a meaningful and ordered sequence for stellar spectra were based on the strength of the Balmer lines; spectral class A came first because these stars had the strongest lines. Secondly (and this was soon realised at Harvard), the strength of not only the Balmer lines but those of any element is as much if not more to do with temperature than abundance or amount of the element itself; this more than anything resulted in the letters of the original Draper classification system which ran from A through Q to be rearranged with some letters even being removed. The tireless efforts of Annie Jump Cannon gave us the spectral sequence which we know and use today. The fact that it is a sequence is very significant; it means that it is not just a 'pigeon holing' of stars into different species but rather shows a steady progression of the nature of the spectra as we go from one end of the sequence to the other. This is born out further by the fact that the sequence does not simply 'jump' say from class A to class F; rather there is a steady transition from A0 to A1 to A2, etc. to A9 which is then followed by F0. F0 is in turn followed by F1, F2, etc. Many stars do have a similar chemical composition and this is why temperature rather than chemistry is the key to the spectral sequence.

Chemistry does have a role to play though; hydrogen is by far the most abundant element coming in at about 75% of all atoms. Next comes helium at about 24% with everything else making up about 1% of the chemical composition of many stars. If an element is very rare then its presence is going to be very difficult to detect in stellar spectra—lithium, the third element in the periodic table is a good example of an element that is so rare that it is virtually not seen in stellar spectra at all. By contrast, fairly common elements which themselves are common products of thermonuclear fusion processes in stars will generally play a more significant role. Common elements include oxygen, nitrogen, carbon as well as helium of course together with metals (as defined by chemists) like calcium, magnesium, sodium and iron. All elements other than hydrogen and helium are referred to as 'metals' by astronomers.

When hydrogen gets hot enough, it becomes ionised and plays no role in the formation of absorption lines. With more complex atoms the situation is much more interesting; in cooler stars they will be electrically neutral and transitions within these neutral atoms will involve the outermost electrons. As temperatures rise, metals will become increasingly ionised losing first one electron and then perhaps more, so transitions now involve electrons from lower energy levels and these transitions and the resulting lines are of course completely different to those from the neutral atoms. Helium because of its relatively large abundance has a significant role to play but because of the energy needed to excite and possibly ionise helium atoms, it is very much restricted to the hot end of the spectral sequence. While lines of many elements gradually put in an appearance as we move along the spectral sequence, molecules of chemical compounds appear when we approach the cool end. Molecules don't produce lines but rather bands in the spectra of cool stars and in fact by the time we reach spectral class M, they are the dominant feature. Finally, there are stars which don't fit into to the main OBAFGKM sequence and the reason they don't is very often to do with chemical composition. So let's have a look at the spectral sequence in the light of what we've learned, particularly about the key role played by temperature.

The Spectral Sequence

The first thing to notice is how the continuum changes as we move along the sequence; not surprisingly for hot stars the continuum peaks in the blue or even the ultraviolet part of the spectrum. This gradually shifts towards longer wavelengths as the temperature lowers and by the time we reach spectral class M the continuum is peaking in the infrared. Even though the spectra of stars have been degraded from the near black body spectrum of their interior layers, this overriding character of a black body spectrum stills remains. In fact it enables professional astronomers to theoretically 'reconstruct' the spectrum that a star would have if indeed it did shine as a perfect black body and to calculate the temperature of such a 'black body star'. This temperature is called the *effective temperature* of the star and is often written as T_{eff} in the literature.

Class O

The hottest class O stars (the very hottest ones are O5) have surface temperatures of around 37,000 K and even by the time we reach the 'cool' end of this class we're still talking temperatures of above 30,000 K. At these temperatures most of the hydrogen in their atmospheres will be ionised. Surprisingly though, lines due to hydrogen are still there even if not very strong. The main reason for this is that the more hydrogen atoms that get ionised, the more free electrons get 'pumped' into the surroundings. Eventually there is a kind of 'saturation' effect (the correct scientific term is *equilibrium*) and electrons can re-attach themselves to hydrogen nuclei where they absorb line photons before they get thermally re-ionised.

The strongest lines in these stars' spectra are those due to ionised helium. Ionised helium or HeII has one electron removed and the result is that it behaves in many ways like a hydrogen atom. The big difference is that the extra positive charge in the nucleus means that the remaining electron is much more tightly bound than is the case for hydrogen. A curious result is that an electron in the $n = 4$ level has roughly the same energy as the electron in the $n = 2$ level for hydrogen and so transitions which start from level 4 in helium produce lines with wavelengths very similar to the Balmer lines for hydrogen.

Other lines in class O stars include those due to doubly and triply ionised oxygen and nitrogen but overall the general appearance of a class O spectrum is relatively 'clean'. Well known examples of class O stars are λ Orionis and S Monocerotis.

Class B

Pick almost any bright star in Orion (Betelgeuse excepted of course) and the chances are that it's of class B; this of course includes Rigel. Class B embraces a relatively large range of temperatures from around 28,000 K at B0 dropping to about 11,000 K at B9. One result is that helium in their atmospheres is no longer thermally ionised. Lines from neutral helium replace those of ionised helium; the result is a more complex looking spectrum because as described in Chapter 3, neutral helium produces two virtually separate groups of spectral series; these are the singlet series which involve transitions between single energy levels and the triplet series whose lines are essentially three very closely spaced lines. They result from transitions between levels which are split into three, though as mentioned in Chapter 3, the lines are so close together that they will simply appear as one line. Another thing to say about the lines from HeI is that they fade as we move through the subdivisions of class B. This is because optical lines result from transitions from the $n = 2$ level and above and it takes relatively high temperatures to thermally excite helium to this level.

Again, as we move through class B, the Balmer lines begin to strengthen. The lowering temperature simply means that progressively less of the hydrogen atoms are ionised and the $n = 2$ level becomes increasingly populated.

Class A

As described above the $n = 2$ level in hydrogen is as populated as it gets at about 10,000 K; the temperature of class A0 stars. As we progress through the class the Balmer lines inevitably fade but interesting things are starting to happen with what the chemists call metals—sodium, calcium, etc.

One of the interesting things about going through the periodic table is that every so often you come to what's called a 'noble gas'. In a sense helium is the first of these which is why chemists place it in the same column as the others. Helium of course has two electrons which if temperatures are relatively low, reside either in the 1s level (if the electrons spins are opposite) or in the 1s level and one of the $n = 2$ levels if the spins are the same. Successive elements fill the $n = 2$ level and become ever more tightly bound to a nucleus with progressively increasing electric charge. By the time we reach neon, the first of the true noble gases we have a tightly bound closed shell of eight electrons in the $n = 2$ level. These electrons do a good job of screening the nuclear charge. After this elements like sodium and magnesium have outer electrons which benefit form this nuclear screening and are much less tightly bound. They thus become ripe for ionisation at moderate temperatures. As we progress through the next few elements, outer electrons become increasingly bound until the next closed shell forms with argon. Once again the pattern repeats with potassium and calcium having easily removed outer electrons.

So the middle section of the spectral sequence is really the metal zone. With the higher temperatures of class A, metals like sodium and calcium are easily ionised. However, in the case of one valence electron metals like sodium, removal of this outer electron leaves an electron structure rather like that of neon; i.e. a tightly bound shell whose individual electrons would require very high energies for their removal. This is also the case with doubly ionised two valence electron metals like calcium. Instead, lines due to singly ionised metals such as CaII begin to appear.

The classic A type stars are Sirius, Vega and Altair.

Class F

These stars span a relatively narrow temperature range from around 7000 K to around 6000 K. Lines due to singly ionised metals are still there and adding to the overall complexity of the spectra are lines from neutral metals. So, for example the doublet series from sodium which includes the D line make their appearance.

The Balmer lines are fading with the drop in temperature; electrons are dropping down into the $n = 1$ level and in consequence the Balmer jump also declines.

A famous F5 star is Procyon.

Class G

The most famous class G star is of course our own sun with a surface temperature of around 5800 K. Class G spectra are dominated at optical wavelengths by line series from both neutral and ionised metals. The strongest lines in the Sun's spectrum are Fraunhofer's H and K lines; a doublet line from singly ionised calcium.

Class K

By this stage temperatures have fallen to around 4000 K; too cool for metals to become thermally ionised and so the spectra of these stars are dominated by lines from neutral metals. At these temperatures some chemical compounds remain intact and so evidence for the presence of titanium oxide (TiO) shows itself, not in the form of lines but of shaded bands. The most famous K star is Arcturus.

Class M

With temperatures around 3000 K the cool end of the spectral sequence is dominated by shaded or fluted bands due to molecules. As variable star observers know most red stars are variable; red dwarfs show flare activity whereas red giants show variability due to pulsation and indeed these stars are one of the main areas of amateur variable star astronomy. Near maximum brightness, red giant variables can show emission lines from the Balmer series. The whole of the next chapter is devoted to explaining the mysteries of molecular spectra.

Others

There are some stars whose classification 'hangs out' on the side of the main spectral sequence, essentially because of differences in chemistry; these are the carbon stars and the zirconium stars. 'He's not going to mention Wolf-Rayet stars' I hear you say; these are a bit special and we'll deal with them later.

Class S

I remember many year ago when I first got interested in observing variable stars, one of the first telescopic variables which I managed to locate and make visual estimates of was R Andromedae. This star is actually the first star listed in the General Catalogue of Variable Stars (GCVS) and it is a spectral class S star. This means that instead of its spectrum (which would otherwise be very much like that of a typical red giant) being dominated by bands of titanium oxide, there are instead bands of zirconium oxide (ZrO).

Class C (Formerly R and N)

Class M and S do have one thing in common, a relatively high chemical abundance of oxygen. With some cool stars though carbon is more abundant and instead of oxygen combining with metals to produce bands of metal oxides, carbon combines with hydrogen and nitrogen to form simple carbon molecules which are seen as bands of CN, CH, etc.

Class L and T

These are 'newcomers' in the spectral sequence and are used to classify very low luminosity stars which are cooler than M (red) dwarfs.

Line Broadening

Clearly all lines in stellar spectra will be subject to natural line broadening and also to a greater or lesser degree of thermal broadening which is commensurate with their temperatures. Another interesting feature which distinguishes giants and supergiants from normal dwarf stars is the presence or lack of pressure broadening. Dwarf stars have relatively dense atmospheres and as a result their lines show well-developed wings which result from pressure broadening. By contrast, lines from giants and supergiants which show no pressure broadening show evidence of enhanced line cores due to turbulence in their vast convective atmospheres.

Rotating Stars

Anyone who has systematically observed sunspots knows that the Sun rotates about once per month and it follows that one would expect other stars to rotate too. If a spectroscope were directed towards the Sun's limb which is rotating away from us then clearly lines in the spectrum from this limb would be red shifted by the Doppler effect. Conversely, spectral lines which come from the Sun's other limb would show a blue shift and finally those which come from the central part of the Sun's disk would show no shift at all because they're moving across our line of sight.

With stars the situation is a little different because the star cannot be seen as a disk, so one can't point a spectroscope at one of the star's limbs. Nonetheless the star *is* a disk and its spectral lines are a composite of light coming from the approaching and receding limbs as well as from the central regions plus all other parts in between. The result is that a line is spread out or broadened by rotation and what's more, because the line itself is the sum total of absorption at that particular wavelength, spreading the line out in this way has the effect of weakening the line compared to how it would be in the absence of rotation. For an absorption line the result is to make the line less deep and for an emission line less tall, so an absorption line will have a washed out or less contrasting appearance, while an emission line will appear more rounded and less bright.

Lines have been observed particularly in hotter stars, which have this appearance but the real indication that this is due to rotation is that the degree of broadening is in direct proportion to the wavelength of the line, i.e. the longer wavelength lines are wider. This comes from the basic equation for the Doppler effect which states that the change in wavelength is directly proportional to the wavelength itself. It is also possible using a little trigonometry to show that the line profile is somewhat dish shaped or ellipse shaped for a star of uniform brightness across the disk, i.e. no limb darkening. When limb darkening is present the centre of the line profile deepens relative to the wings and the profile becomes parabolic. As described in Chapter 5 for other line broadening mechanisms, these rotation line profiles would be convolved with the usual Doppler broadened profile. One final thing; the effects of rotation will only appear if the rotation rate of the star is sufficiently high and also if its axis of rotation lies close to the plane of the sky; a star seen pole on will show no rotation effect.

Imagine an alien astronomical spectroscopist observing our sun from many light years away; let's see what the effect of its rotation on its spectral lines would be. The radius of the Sun is 6.96×10^8 m and hence its circumference is 4.37×10^9 m. Its equatorial rotation period is approximately 24.9 days; this equals 2.15×10^6 s. A point on the surface has to travel one complete circumference in one rotation period; so the equatorial rotation velocity is equal to 4.37×10^9 divided by 2.15×10^6 which equals 2.03 km/s. If you plug this into the simple Doppler shift formula—Eq. (4.1)— the wavelength shift at λ6563 is equal to about 0.044 Å. This means that at best our alien astronomer would see the Hα line broadened by about 0.09 Å due to the Sun's rotation. He'd probably quickly turn his attention (and expensive telescope time) to observing spectral class A stars which are known to often have rapid rotation rates.

Spectral Snapshots

Amateur astronomers who have done CCD photometry or photoelectric photometry know that to do the job properly you need to make observations through the so-called standard Johnson filters. One of the main things which distinguishes one star in the spectral sequence from another is the strength or brightness of the continuum in different regions of the spectrum; for example the hottest stars will peak in the ultraviolet and will still be bright at blue wavelengths. By contrast, the coolest stars will peak in the infrared with a relatively bright continuum at red wavelengths (not withstanding the presence of many dark absorption bands). A relatively straightforward way to sample the spectrum of a star is to determine its magnitude in different wavelength regions.

To do this the light from the star is allowed to shine through one of the Johnson filters before it either falls onto the CCD or the photometer sensor.

The Johnson filters for the optical spectrum are designated U, B and V which stand for 'ultraviolet', 'blue' and 'visual', respectively. The visual filter has maximum transmission at $\lambda 5500$ and covers the region of the spectrum to which the human eye is most sensitive, i.e. the yellow-green region. The parts of the spectrum covered by the ultraviolet and blue filters are centred at $\lambda 3600$ and $\lambda 4500$ respectively and they sit either side of the Balmer jump. The end result of allowing the light from the star to pass through one of these filters is a magnitude in exactly the same way as is 'ordinary' apparent magnitude and it is measured in exactly the same way. So for instance, a hot blue star will be brighter in the blue part of the spectrum than a cooler yellow star, hence a class O star will have a brighter blue or B magnitude than a class G star. Indeed, the class O star's B magnitude will be brighter than its own Visual or V magnitude but by contrast the G star will have a brighter V magnitude and a fainter B magnitude.

From these *colour magnitudes*, astronomers produce magnitude differences; $U-B$ and $B-V$ for a given star. Each of these magnitude differences is called a *colour index*. The very magnitude system itself is formulated so that stars of spectral class A0 have a $B-V$ colour index of 0.0 and so $B-V$ for O and B stars will be negative whereas that for spectral classes later than A0 will be positive. Clearly, $B-V$ is closely related to spectral class but what does this have to do with *your* spectra?

I said at the beginning that starlight has a difficult journey, dodging amongst other things interstellar dust. Some of that light doesn't make it through the dust though and blue light is particularly prone to being 'knocked out' by the interstellar medium. Stars which are more distant will inevitably shine through more intervening material and hot stars in particular (class O and class B) will have their copious supplies of blue photons diminished. This in turn will affect their B magnitudes which in their turn affect the $B-V$ colour index. B will lose out and become fainter so its numerical value will increase just as it does for ordinary apparent magnitude. This makes $B-V$ more positive—it's as if our hot blue stars are behaving like cooler red stars. If the star were shining through clear empty space, it would of course have its appropriate value for $B-V$; let's call this $(B-V)_0$. Because of interstellar absorption, however, what we observe is a more positive value which we can just call $(B-V)$. If we subtract $(B-V)_0$ from $(B-V)$ we get a number which is always positive. This is written $E(B-V)$ and is called the *colour excess*. So interstellar absorption means that if you take spectra of more distant stars, they will effectively appear redder than they should for their spectral class and this effect is not surprisingly is called *interstellar reddening*. If you think about it though this is the result of blue light being removed from a star's spectrum and so the term interstellar de-blueing might be more appropriate, though this doesn't role off the tongue so well.

A Word or Two About the Herzsprung–Russell Diagram

Just as soon as any amateur astronomer gets to know about the spectral sequence, he or she gets to know about the most famous diagram in all of stellar astronomy—the Herzsprung–Russell (HR) diagram. Indeed, our first acquaintance is likely to show the spectral sequence as forming the horizontal axis of the diagram with the hot stars to the left and the cool stars to the right. The vertical axis shows some quantity which measures the luminosity or true power output of the star and this is usually the absolute magnitude. However, determining the spectral class of a star (let alone a large population of stars) is not a trivial task. The important thing is that the horizontal axis plots stars according to some quantity which represents their temperatures. As we have seen, the spectral sequence is a temperature sequence hence its use in the HR diagram.

There are two other quantities which can be used to effectively determine the temperature of a star; one of these is the effective temperature and indeed this is the quantity which would probably be used by a theoretical astrophysicist who was interested in developing models of stellar atmospheres. Such an HR diagram would then be known as a *theoretical HR diagram*. A much more common practise though is to use the $B-V$ colour index; this 'spectrum snapshot' which of course is done (relatively easily) using photometry rather than spectroscopy leads to what is known as an *observational HR diagram*.

Summary

- Photoionisation of hydrogen is the chief cause of continuum absorption in hot stars.
- The negative hydrogen ion is the chief source of continuum absorption in cooler stars.
- Line absorption (i.e. the formation of absorption lines) is as much if not more to do with temperature than chemical composition.
- The Harvard spectral sequence is in consequence based on the temperatures of stellar atmospheres.
- Stellar photometry through standard filters provides a relatively straightforward way to determine stellar temperatures using colour indices.
- Interstellar absorption removes blue wavelength light from a star's spectrum affecting its colour indices and making it appear redder.

Cool but not Smooth—The Molecular Spectra of Red Stars

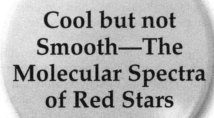

For my PhD I had to model the Hα line profile in the spectra of symbiotic stars; these are interacting binary stars in which one of the components is a cool giant often of spectral class M. The continuum in the vicinity of the Hα line was seemingly very 'noisy' in one or two of the spectra which I was using and I remember someone at the time commenting that perhaps all the myriad spikes and dips weren't in fact instrumental noise but the titanium oxide (TiO) bands in the spectrum of the red giant. I recall even then thinking 'is the continuous spectrum of a red giant such terra incognita that one cannot distinguish the absorption bands from instrumental noise?' It's true that in most books which list the spectral classes, the section on class M can sometimes be almost dismissive with phrases like 'dominated by bands of titanium oxide and other molecules'. However, if you dig deep enough in the literature you'll certainly find that much is known about the patterns of these mysterious bands to the extent that even bands due to different isotopes of titanium have been identified; even so it will probably be a struggle to identify individual bands in your spectra.

A good start though might be to understand why molecules like titanium oxide produce these bands rather than the relatively simple lines of atomic spectra. This is mainly what this chapter is about, but another reason for doing it is that cool giants more often than not tend to be variable stars; either irregular, semi-regular or Mira-type variables. As such they are of course of huge importance to both amateur and professional variable star observers. By contrast, you could get the impression that red dwarfs, i.e. cool main sequence stars, are the 'poor man's end' of stellar astronomy. Their claim to fame though is when they spectacularly and very rapidly brighten as flare stars. They too, like their big brothers have very complex spectra which include bands due to molecules.

Stellar Atmosphere Versus the Chemistry Lab

Temperatures in the outer layers of class M (and class S for giants) stars can drop below the 2000 K mark and even at typical temperatures of 2500 to 3000 K it's possible for certain molecules to survive intact against thermal dissociation. What's more, unlike conditions in a typical chemistry lab experiment, densities, particularly in the outer layers of red giants and even in red dwarfs are relatively low; these low densities further help to prevent thermal dissociation of molecules by lowering the frequency of inter-molecular collisions.

Cool giant stars have exceedingly complex looking spectra due largely to the presence of seemingly countless absorption bands due to molecules. These bands are often due to molecules of titanium oxide (TiO) or maybe zirconium oxide (ZrO) which can survive in the relatively low temperatures of cool giant stellar atmospheres. There are also bands due to molecules like carbon monoxide (CO) and even water (H_2O) and then there are also odd things like CN and NH, etc. Besides hydrogen and helium, elements which are relatively common in the outer layers of most stars are carbon (C), nitrogen (N) and oxygen (O). Atoms of these elements can bond together to produce structures which you wouldn't find in a chemistry laboratory. In the chemistry lab they are called *radicals* and the reason they don't exist separately is again due to the fact that densities in test tubes are relatively high and frequent collisions mean that they either get dissociated or combine with other radicals to form complete molecules. In the atmospheres of red giants though, the much lower densities ensure that collisions are infrequent; these radicals are actually physically stable if left alone and so they survive to produce their own molecular spectra.

The molecular bands themselves are each made up of a large number of closely spaced individual lines which often crowd together on one side to form what is called *a band head*. Large numbers of bands stretching across the visible spectrum amount to literally millions of individual lines. This is clearly in stark contrast to the relatively simple spectral series formed by electron transitions in atoms and the key to understanding why this is so lies in learning what a molecule can do that atoms can't.

The Things That Molecules Do

A molecule consists of two or more atoms chemically bonded together and to keep things simple, we'll stick to two-atom or *diatomic* molecules. Without going too much into chemistry, the bond may be *ionic* which means that one atom has effectively lost an electron leaving it as a positively charged ion. This electron has in turn effectively been grabbed by the other atom turning it into a negative ion; the two ions are then bound by their electrostatic attraction. Ionic bonds work best with atoms which have one or two relatively loosely bound electrons surrounding a closed shell and those which are just one or two electrons short of a closed shell. The classic case is sodium

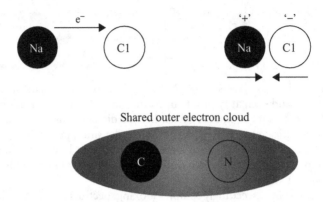

Shared outer electron cloud

Figure 7.1. The two basic chemical bonds. At the top, ionic bonding is caused by electrostatic attraction between two atoms (here sodium (Na) and chlorine (Cl)); one of which has lost an outer electron (sodium in this case). This electron in turn has been captured by the other atom. At the bottom, covalent bonding effectively results from atoms sharing their outer electrons; covalent bonds are usually weaker than ionic bonds.

which has a single valence electron outside of a neon closed-shell structure and chlorine which is one electron short of a closed-shell argon structure. The ionic bond between these two makes a molecule of sodium chloride of course. Elements like calcium and magnesium with two outer valence electrons also make good ionic bonds as do most elements which a chemist would describe as 'metals'.

The other main type of bond is a *covalent* bond; I remember at high school they wouldn't tell us all the details about this kind of bond because its explanation only came with the aid of quantum mechanics and what's worse the explanation as such came in the form of certain terms in the mathematics. Basically though the atoms share one or more pairs of electrons; a possible astronomical analogy might be a close binary star with a common envelope and a culinary analogy might be an egg with a double yolk where the white of the egg represents the shared electron cloud of the molecule. This is crude I know but it'll do for our purposes.

With both kinds of bond it's possible that one of the two atomic nuclei doesn't get shielded by the electrons as well as the other or looked at another way, the outer electrons spend more of their time around one of the nuclei than the other. For a diatomic molecule, which we can think of as a tiny dumbbell, the result is that one end has a slight excess of positive charge and the other a slight excess of negative charge. The molecule is thus said to be slightly polarised and becomes an *electric dipole* with what's known as a *dipole moment*. This dipole moment means that just as a see-saw can be turned on its axis by lifting one end, a dipole can be turned on its axis by applying an outside electric field and what better external field than the electric field part of an incoming electromagnetic wave. Light can turn molecules and make them rotate!

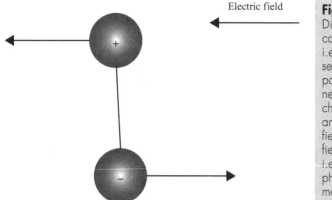

Electric field

Figure 7.2.
Diatomic molecules can be polarised, i.e. there is a slight separation of positive and negative electric charge. This enables an external electric field (e.g. the electric field of a light wave, i.e. an incoming photon) to set the molecule rotating.

Rotation

Molecules can rotate 'end over end' as it were but they can't spin around the axis joining the two atoms as shown in Fig. 7.3.

As with everything it takes energy to turn a molecule. If the energy is that of light, then that energy is lost or absorbed just as it can be when light encounters atoms. Molecules are tiny and are thus subject to the laws of quantum mechanics rather than the 'laws of see-saws'; a good way to think of this is that a given molecule can only rotate at certain fixed discrete rates or frequencies and it cannot rotate at any frequency in between. Each rotation frequency corresponds to the molecule having a different amount of rotational energy; the higher the rotation rate or frequency, the higher the energy. We can thus think of these rotation frequencies as *rotation levels* for the molecule; the faster it rotates the greater the energy and the higher the level.

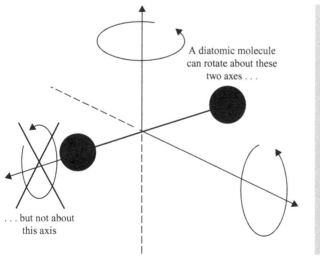

A diatomic molecule can rotate about these two axes . . .

. . . but not about this axis

Figure 7.3. A diatomic molecule can rotate end over end but not about the axis joining the two atoms.

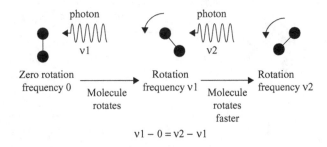

$$v1 - 0 = v2 - v1$$

Figure 7.4. Photons of successively higher frequencies (i.e. energies) are needed in order to get the molecule rotating faster. The frequency difference between the quantum mechanically allowed rotation rates are the same.

Another feature of the quantum mechanics of rotating molecules is that they can undergo *rotational transitions* by either absorbing or emitting a photon; however they can only 'jump' from one rotation level to another, one level at a time; this is in contrast to electrons in atoms which can undergo upward or downward transitions across several levels at once. For a molecule to go from its ground state, i.e. zero rotation to the first rotation level, it must absorb a photon of frequency equal to that of the first rotation level. To get to the next level it must in turn absorb a photon of frequency equal to the next higher level and so on; so getting a molecule to rotate faster means that the molecule has to absorb a succession of photons of increasing frequency (i.e. shorter wavelength).

If we place the usual vast population of molecules of a given kind in an electromagnetic radiation field, then the molecules are going to start to rotate. To get them rotating at the first frequency requires photons of the same frequency and so these

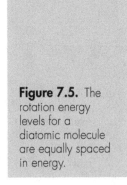

Figure 7.5. The rotation energy levels for a diatomic molecule are equally spaced in energy.

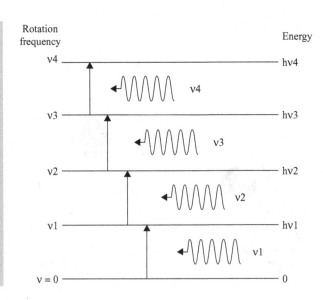

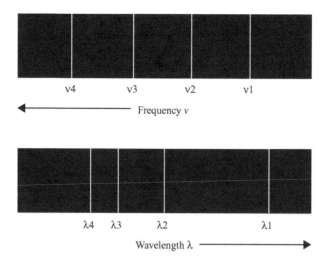

Figure 7.6. A spectrum of pure rotation lines, if plotted in terms of frequency rather than wavelength would result in the lines being equally spaced. Plotted in terms of wavelength, however, the lines would get closer together towards shorter wavelengths. In reality, the lines would be much closer together than is depicted here.

photons get absorbed; matching photons of higher frequency will raise the molecules to the next rotation level and in doing so are absorbed too. This process results in photons at a series of frequencies which match those of the rotation rates of the molecules being absorbed and producing a set of absorption lines. Unlike the energy levels in atoms, the rotation levels for diatomic molecules are equally spaced in energy and hence frequency, so a spectrum of these rotation lines if plotted as intensity against frequency rather than wavelength would show the lines to be equally spaced. However, if plotted in terms of wavelength which is the norm for astronomical spectra, the lines would get closer to each other towards the shorter wavelengths.

So far, quite simple; a population of molecules which only performed rotation transitions would produce a very simple spectrum indeed—a simple set of lines. At this stage though things are rather academic because it takes relatively little energy to get the molecules rotating and so pure rotation lines would only be seen in the far infrared or the sub-millimetre region of the spectrum.

Vibration

Now for some fun! Our little dipoles besides being made to rotate by the electric field of an electromagnetic wave can also be 'squashed' slightly along the line joining the two atoms. The positive end of the dipole gets pushed slightly by the electric field, while the negative end gets pulled; this brings the two (positively charged) atomic nuclei closer together and the two atoms immediately push back. This sets up a tiny vibration along the line joining the two atoms and yes, you've guessed it, the frequency of the vibration is again subject to the laws of quantum mechanics. A vibrating string on a violin or a guitar has a fundamental note plus a series of discrete harmonics or overtones and

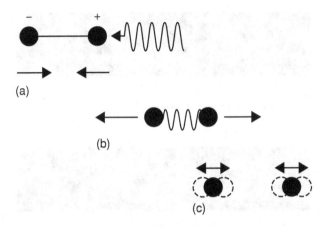

Figure 7.7. An incoming photon can 'squash' a diatomic molecule along the line joining the two atoms as shown in 'a'. The two positively charged atomic nuclei briefly get closer and immediately push each other apart as shown in 'b'. This sets up a vibration just like two weights at the ends of a tiny spring as shown in 'c'.

so do diatomic molecules. Each of these vibration overtones or *vibration levels* as the physicists prefer to call them, gives the molecule a slightly different amount of energy and once again within a vast population, molecules would exist in various vibration states or levels; there is even a separate quantum number to label these vibration energy levels which not surprisingly is called the *vibration quantum number n_v*.

At first we might think that a molecule could undergo a transition from one vibration level to another as a result say of absorbing a photon of appropriate wavelength and then within a vast population of molecules undergoing various vibration transitions, the result would be a series of vibrational absorption lines. This doesn't happen however because when a vibration transition does takes place, it is always accompanied by a rotation transition which as described above involves a jump of one and only one rotation level. A large number of molecules which are all undergoing the same vibration transition will be in different rotation levels and so will produce a whole range of rotation transitions; in fact it's easy to show that if a molecule had for example five rotation states including zero rotation then the total possible number of different transitions between two vibration levels is 8, i.e. $(5 - 1) \times 2$. The transition from zero to zero rotation is not allowed. Fig. 7.9 shows the different possible transitions and clearly it's exactly as if each vibration level is split into several sublevels, so that what would otherwise be a single line resulting from one transition between two vibration levels, now becomes split into several lines by the 'rotation sublevels'.

All of these *vibration rotation transitions* involve slightly different energies and so the result is a series of closely spaced lines which converge together at the short wavelength end as shown in Fig. 7.10. Though vibration levels involve more energy than rotation levels, a set of *vibration rotation lines* by itself would still only be seen in the near infrared and a final point is of course that because you can't have a vibration transition without a rotation transition, pure vibration lines are never seen. We now begin to see where the complexity of molecular spectra comes from but there's still one more layer to add to this molecular spectrum cake.

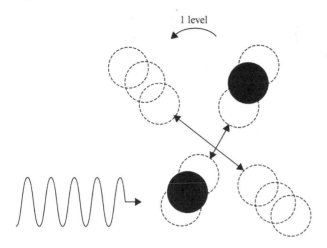

1 level

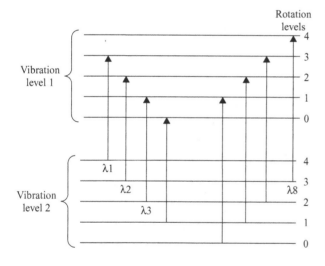

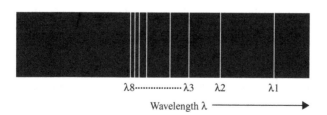

Figure 7.10. A single vibration rotation series (i.e. between two vibration levels) would produce a series of closely spaced lines which converge towards shorter wavelengths.

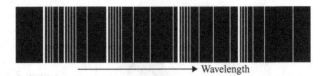

Wavelength

Figure 7.11. A stylised (and very much simplified) electron vibration rotation transition series (i.e. a molecular band). The thicker lines represent the head of each vibration rotation series, which themselves also get closer towards shorter wavelengths to form the molecular band head.

Electron Transitions

Just as atoms make their spectra from electron transitions, molecules can undergo electron transitions too and this takes vibration rotation lines into the visible spectrum league. Again, as with atoms, electron transitions by themselves would produce a simple series of lines but all the molecules within a large population which are undergoing a given electron transition are in various vibration levels also undergoing vibration transitions together with their accompanying rotational transitions. So each electron energy level is effectively split into a series of vibration levels which in turn are each split into a set of rotation levels. One single line for an electron transition first becomes a series of lines due to the different vibration transitions; these converge towards the shorter wavelengths. Each of these vibration lines in turn splits into a series of rotation lines again converging towards the short wavelength end. The overall result is a series of many lines which crowd together towards shorter wavelengths—a *molecular band* with a *band head* at the short wavelength end. Fig. 7.11 shows a very simplified stylised version of a single electron transition split into five vibration levels which in turn are split into several rotation levels. In reality, there would be many more individual lines here and the crowding effect would be much greater.

Add to this band those produced by other electron transitions from the same kind of molecule and the result is many overlapping bands making the visible spectrum of a cool red star a sight to behold. Oh and yes of course, somewhere in there are lines due to atoms too. Well there we are; we've made a molecular band, a series of molecular bands and a whole molecular spectrum just from one type of molecule; add extra molecules to the recipe and the saga of a cool star spectrum is complete. All of this will probably not help you to sort out the 'mess' on your spectra but I hope you now understand better where this mess comes from.

Summary

- Molecules exist in the relatively low temperatures of cool stars' atmospheres.
- Diatomic molecules can rotate at a set of fixed rotation rates or frequencies.
- A rotation transition occurs when a molecule absorbs a photon of frequency equal to that of the next higher rotation rate.

- Rotation transitions involve very low energies so pure rotation spectra are not seen in the visible part of the spectrum.
- Diatomic molecules can vibrate along the line joining the two atoms, at fixed vibration frequencies.
- A vibration transition is always accompanied by a rotation transition.
- Vibration rotation transitions spectra can sometimes be seen in the infrared part of the spectrum.
- A transition between two electron energy levels effectively splits into a series of vibration transitions each of which in turn splits into several rotation transitions.
- The result is a molecular band in which large numbers of individual lines crowd together to form a band head at the short wavelength end.

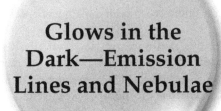

Glows in the Dark—Emission Lines and Nebulae

So far we've been dealing mainly with the continuum and absorption lines in the spectra of normal stars. It's time we looked at emission lines; some stars like Wolf-Rayet stars and symbiotic stars together with Mira variables do produce emission lines in their spectra. However, the best place to introduce our emission line studies is gaseous nebulae and planetary nebulae; their spectra consist almost exclusively of emission lines. They have also been for many years now, great favourites with deep sky observers and so there is inevitably going to be great interest among amateur spectroscopists who want to know how these beautiful objects work.

What Comes Down Must First Go Up

Emission of photons by atoms results from downward bound–bound transitions. Before something can come down, in this case an electron, it must first go up. We've already seen how thermal excitation in stellar atmospheres can populate higher energy levels in atoms, making them suitable for producing absorption lines for a given series. In the case of planetary and gaseous nebulae however the surrounding gas is relatively cold and so the chances are that atoms which make up the gas will be in their ground state, i.e. any electrons will be in the lowest energy levels. However, one thing that both planetary and gaseous nebulae have in common is hot stars; planetary nebulae have very hot central stars which are essentially the exposed reactor cores of former red giants and gaseous nebulae often have young spectral class O stars embedded within them.

In Chapter 2 we talked about the black body spectrum and noted that as a body gets hotter, the wavelength at which it produces maximum emission of energy gets shorter. The physicist Wien in fact produced an extremely simple formula for calculating this peak emission wavelength if you know the temperature of the body. This can be simply written as

$$\lambda_{max} = 28978200/T \qquad (8.1)$$

T is the temperature of the body in Kelvin and λ_{max} is the wavelength in angstroms which we're looking for. We already know that a star doesn't radiate as a perfect black body but assuming that it does is not too bad an approximation. If we know the effective temperature of a star, we can easily calculate the wavelength of maximum energy emission from the star using Eq. (8.1) which by the way, is known as *Wien's displacement law*.

Let's try it out with some typical stars starting at the cool end with a red giant which has a typical effective temperature of around 3000 K. Eq. (8.1) will give us 28,978,200/3000 which equals 9659 Å. This is well into the infrared which is typical for cool red stars. Now let's try the sun at 5800 K; this will give λ_{max} equal to 4996 Å which lies in the green part of the visible spectrum. It's perhaps no surprise that human eyes have evolved to be most sensitive to green light. A spectral class O star such as would typically be found within a gaseous nebula has by Eq. (8.1), maximum energy emission at 783 Å which is well into the ultraviolet and finally the central star of a planetary nebula, shining at a temperature of perhaps 100,000 K maximally blasts out photons at λ280 in the extreme ultraviolet.

So the hot stars within gaseous nebulae and planetary nebulae produce large supplies of high-energy ultraviolet photons. These high-energy photons are the 'stuff' that makes the electrons in the surrounding gas atoms 'go up'. Let's check this to see what kinds of photons are indeed needed to make electrons actually leave the surrounding atoms, i.e. to ionise them. If a star produces photons which can ionise some of the surrounding gas atoms, then it will certainly produce photons which are capable of merely exciting others.

We'll start with hydrogen; both gaseous nebulae and planetary nebulae contain large amounts of hydrogen. So let's see what happens when hydrogen atoms in the ground state get blasted with high-energy ultraviolet photons. Remember in Chapter 3 I gave a simple equation which converts electron volts (ionisation potentials are almost always given in these units) to a wavelength in angstroms. This equation simply says; divide 1.24033×10^4 (you could write this as 12,403.3 if you like) by the number of electron volts to get the corresponding wavelength. The ionisation potential for hydrogen in the ground state is 13.598 eV and dividing this into 12,403.3 gives an equivalent wavelength of 912 Å. This means that photons with a wavelength equal to or less than this will ionise hydrogen atoms and clearly both planetary nebula central stars and spectral class O stars can do this.

Another common element in nebulae is helium and this has two electrons. The ionisation potential for removing the first electron is 24.587 eV and that for removal of the second electron is 54.416 eV. These ionisation potentials correspond to photons at λ504 and λ228. Even for these short wavelengths a class O star will produce sufficient photons to singly ionise helium and a planetary nebula star will produce plenty of photons which are capable of producing doubly ionised helium or HeIII.

Other common elements found in nebulae are carbon nitrogen and oxygen and with these elements there's scope for high levels of ionisation. Photons at λ192 will remove all four valence electrons from carbon atoms; 160 Å will do the same job on nitrogen and also on atoms of oxygen. So a planetary nebula will inevitably contain a lot of highly ionised atoms; class O stars in gaseous nebulae won't do quite so much 'damage' but there will still be a lot of ionisation going on in places like the Orion Nebula. So what then?

Recombination

The ionising effect of high-energy photons produces in a surrounding nebula a 'sea' of ions and electrons; physicists call this a plasma. Things don't stay that way though; the unlike charges of ions and electrons inevitably cause electrons to be recaptured onto ions and then something wonderful happens. Hot stars supply the big bucks; the hundred dollar bills in the form of high-energy photons which ionise the atoms, but when an electron meets up with an ionised atom it often starts on a higher energy level. It then 'cascades' (and the term for this process really is cascade) down the energy levels maybe only one or two levels at a time. Each of these downward bound–bound

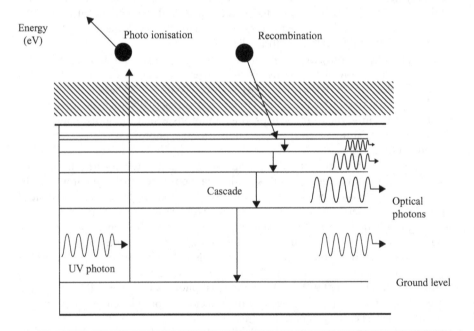

Figure 8.1. The basic process which produces the emission line spectra of nebulae; high-energy ultraviolet photons from hot stars ionise atoms within the nebula. Electrons are subsequently recaptured by atoms, often on higher energy levels. A captured electron then cascades down through lower levels emitting a series of optical emission lines.

transitions might produce emission of a visible light photon which can contribute to an emission line in the visible part of the spectrum. This cascading process takes place very rapidly in perhaps only 10^{-8} s; the atom is then re-primed and ready for another incoming high-energy photon.

The process of recapture of an electron by an ion is perhaps not surprisingly called *recombination* and spectral lines which result from the ensuing cascade process are called *recombination lines*. Clearly even with just the five main nebula constituents of hydrogen, helium, carbon, nitrogen and oxygen, a very large number of line series are possible when you consider the varying degrees of ionisation which are likely to be present. The ionisation recombination cascade process depends initially on the supply of high-energy ultraviolet photons from a hot star but as the process takes place throughout the nebula, more and more of these high-energy photons get converted into (larger numbers of) lower energy photons. In other words, high-energy photons get degraded into low energy ones and what's more as we move outwards through the nebula away from the hot star, this degradation increases.

Photon Degrading and Recycling

Let's start near the surface of the hot star; a veritable blast of high-energy photons almost immediately encounters gas atoms in the surrounding nebula. Suppose that for a very brief instant, an oxygen atom has all of its eight electrons present; in the next instant the 'hail' of photons will strip four of its electrons away to leave an OV atom. It's taken a 160 Å photon to remove the fourth electron and when that electron recombines with another OV atom it's possible that we might get that photon back. What's more likely to happen though as a result of cascading is that it will be broken down into several longer wavelength photons; however one of these photons could still be of pretty high energy. Let's keep this in mind but note most of all that relatively close to the hot star is where the highest energy photons get degraded. So the highest degrees of ionisation and any spectral series that result will take place close to the hot star. This means also of course that the supply of the highest energy photons becomes exhausted first and it turns out that once exhaustion 'sets in', it happens very rapidly; the photons don't 'fizzle out' gradually. The result is that there is a relatively sharp boundary within the nebula inside of which the highest ionisation levels will be found and outside of which they won't be found.

Now back to that pretty high-energy photon which of course will join the countless others that themselves have been produced by downward transitions in highly ionised atoms and those which are coming from the star itself. These 'pretty high energy' photons will cause atoms to have lesser degrees of ionisation such as OIV, OIII, etc. Once again recombination followed by cascade means that they in their turn become degraded into still lower energy photons. So as we move out through the nebula, we see a gradual decline in the population of the highest energy photons and a relative increase in the number of lower energy photons.

Now let's go back to the surface of the hot star; besides the high-energy photons there will of course also be plenty of lower energy ones. These won't have any effect on those atoms and ions which have high ionisation potentials but they will be able to ionise hydrogen atoms provided their wavelength is less than 912 Å. Recombination here can result in Balmer emission lines in the visible part of the spectrum. The difference though is that the supply of these kinds of photons is constantly being replenished by degradation of higher energy photons so they are not confined to the region close to the star. The result is that ionisation of hydrogen can take place more or less throughout the nebula. So summarising, lines which are due to high levels of ionisation come from regions close to the hot star whereas lines requiring less energy are spread more or less throughout the nebula.

The effect of this can be seen by producing a spectrum of a planetary nebula without using a slit in the spectroscope. The result is a series of separate images of the nebula each corresponding to one of the brightest lines in the spectrum. The apparent size of each image depends on the line producing it; the smallest images come from the highest ionisation lines. The ionised zone for a given element or ion is often referred to as the *Strömgren sphere* for that element. The word sphere is perhaps a little inappropriate because many planetary nebulae for example are known to be not spherical but 'hour glass' shaped and gaseous nebulae are generally anything but spherical.

Thick and Thin Nebulae

To ionise hydrogen atoms which are in the ground state (i.e. with their electron in the $n = 1$ level), photons with a wavelength equal to or less than 912 Å are necessary and these will be plentiful in the radiation field within a planetary nebula or a gaseous nebula. These photons make up what is called the *Lyman continuum*. What happens next is interesting and revealing; when recombination takes place the electron will probably start on a higher energy level. It could go straight back down to level 1 of course and this would just give us back our original photon though this photon will almost immediately get absorbed by another atom, so Lyman continuum photons will never escape the nebula.

Those electrons which do recombine on a higher energy level will cascade very rapidly down the lower levels and will eventually reach level $n = 1$ again because this is the 'natural' state for hydrogen at the temperatures which exist within nebulae. The final cascade jump may well be from level $n = 2$ to $n = 1$; this will produce a Lyman α photon at $\lambda 1215$. This photon has nowhere to go; its wavelength is too long to ionise other hydrogen atoms. It can only excite hydrogen atoms back up to the $n = 2$ level which then almost immediately de-excite and our poor Lyman α photon is back on the street as it were. After being scattered like this many times it can and does eventually escape from the nebula, to be observed together with its buddies as a Lyman α emission line in the ultraviolet part of the spectrum.

The final cascade jump may be between the $n = 3$ and $n = 1$ levels; in this case we get a Lyman β photon at $\lambda 1025$. Again this photon can excite a hydrogen atom back up to the $n = 3$ level but this time there is a chance that it will cascade in two jumps; firstly from $n = 3$ to $n = 2$ giving an Hα photon, and secondly from $n = 2$ to $n = 1$ giving a Lyman α photon which can eventually escape. So in the long run, Lyman β photons don't get away but get degraded; the same thing applies to Lyman γ and other photons in the Lyman series. The only thing that can change

Table 8.1 The theoretical Balmer decrement; the relative intensities of the Balmer lines on a scale where Hβ equals 10, are given for case A and case B recombination in nebulae.

	λ(Å)	Case A	Case B
Hβ	4861	10	10
Hγ	4340	5.76	5.1
Hδ	4101	3.74	3.1
etc.	3969	2.55	2.06
	3889	1.82	1.43
	3835	1.36	1.05
	3797	1.05	0.79
	3770	0.81	0.59
	3750	0.65	0.46

this is if the hydrogen in the nebula is thin; i.e. of low density. In this case Lyman line photons can escape. Recombination which takes place in a nebula which is dense enough to prevent Lyman line photons (except for Lyman α) escaping is called *case B recombination*. When the nebula is of low enough density that Lyman photons can escape, the process is called *case A recombination*. For case B recombination the nebula is basically opaque to Lyman line photons which in consequence get absorbed and degraded. Another way that astronomers describe this is to say that the nebula is *optically thick* to Lyman line photons, whereas for case A recombination the nebula is *optically thin*. These terms are used widely in astrophysics to describe the absorbing properties of any optical medium. If a medium strongly absorbs radiation of a given wavelength then it is optically thick at that wavelength; if it's optically thin the radiation gets through.

I remember at high school looking through a laboratory spectroscope at an emission line spectrum of hydrogen. One thing I noticed was that the lines seemed to start off bright at one end of the spectrum but then gradually got progressively fainter. I noticed the same thing again in the physics lab at university but no explanation ever seemed to be offered. It wasn't until years later when reading astrophysics textbooks and research papers that I came across the term *Balmer decrement* and realised that this referred to the decreasing intensity of the Balmer lines as you move away from the Hα line towards the blue end of the spectrum.

The way that the Balmer decrement is usually defined is to give the intensity of the Hβ line a value of 10 and then scale the intensities of the other lines accordingly. The Hα line is often so strong that lines further up the series would have very small values indeed if the Hα line were used as the standard. The set of ratios which comes from this scaling procedure (i.e. Hγ/Hβ, Hδ/Hβ, etc.) is what is actually referred to as the Balmer decrement. In the 1930s, astronomers went to a great deal of trouble to calculate the values of the Balmer decrement ratios for a given situation. The calculations were extremely tedious but the results matched very well the observed Balmer decrement for planetary nebulae provided you assumed that the nebula was dense enough for case B recombination to dominate. Table 8.1 lists the ratios for the Balmer decrement for case A and case B recombination and these are plotted in Fig. 8.2.

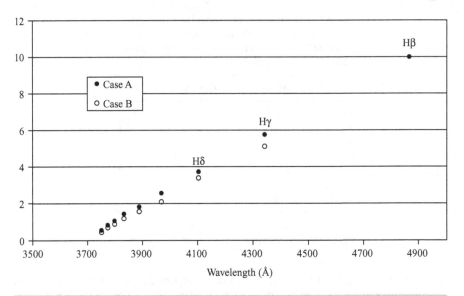

Figure 8.2. A plot of the relative intensities of the Balmer lines in a nebula spectrum for case A and case B recombination. The actual ratios of the line intensities relative to that for the Hβ line are collectively known as the Balmer decrement.

It's possible to see how this decrease in line intensity comes about without going into the maths but if you want to calculate the actual intensity ratios, you'll have to do the tedious calculations yourself. Balmer emission lines result from electron transitions which end on level $n = 2$. Take first an electron in level 3; this can drop down to level 2 to give us an Hα photon. Now take an electron in level 4; this can drop straight down to level 2 to give us an Hβ photon but it may also drop first to level 3 (giving a Paschen α photon in the infrared) and then drop to level 2 to give another Hα photon to add to the collection. Again, starting at level 5 a straight drop will give us Hγ but there are now also the following possible jumps:

Level;

5	→	4	→	3	→	2
5	→	4	→	2		
5	→	3	→	2		

The top sequence gives an α photon of the *Brackett series* plus a Paschen α photon and yet another Hα photon. The middle series gives Brackett α and another Hβ and the bottom series gives Paschen β and another Hα. Try working out for yourself the possible transitions which can take place from level 6. Now imagine the usual vast population of atoms with all these kinds of transitions taking place; it's clear that there will always be a relatively high number of Hα photons produced together with a correspondingly smaller number of Hβ photons and a smaller still number of Hγ photons, etc. So there you have it; the Balmer decrement without the tedious calculations!

Yet More Photon Recycling—Fluorescence

So far we've seen how Lyman photons fail to escape from a sufficiently dense nebula but get absorbed and recycled into other lower energy photons. The fact that nebulae also contain significant amounts of helium, carbon, nitrogen and oxygen mean that a huge variety of electron transitions gives rise to an equally huge variety of photons flying around within the nebula. Some of these too get recycled; the most famous example of this is the *Bowen fluorescence mechanism* after the British astronomer I.S. Bowen. This is restricted to planetary nebulae and also active galactic nuclei (AGNs) because the process starts with doubly ionised helium or HeIII and class O stars in galactic nebulae are not hot enough to produce this. Recombination of HeIII gives rise to emission lines from HeII and the final cascade drop takes the electron from the $n = 2$ level to the $n = 1$ level. The resulting line for this transition has a wavelength of about 304 Å and incidentally the transition which takes an electron between the ground state and the next level up in any atom produces what's called a *resonance line*. By a pure coincidence, the wavelength of this resonance line for HeII almost exactly matches that of a photon which will take an electron in the 2p level of doubly ionised oxygen OIII up to the 3d level (note here the l selection rule is obeyed). Exact matching doesn't matter here because motion of the atoms within the nebula mean that once again our old friend the Doppler effect comes into play ensuring that an exact match can and does happen. So these photons get absorbed. They can then cascade down through the sublevels of the $n = 3$ level to produce a series of lines in the ultraviolet. For the amateur spectroscopist however the important bit is the final drop from the 3s to the 2p level; this emits a photon at about $\lambda 374$ and by yet another coincidence this photon can excite an atom of doubly ionised nitrogen NIII from the 2p to the 3d level. This time however the transitions through the $n = 3$ sublevels together with the final drop from 3s to 2p produce a series of lines in the approximate range $\lambda\lambda 4100$ to 4600 which of course lie in the blue part of the visible spectrum.

Forbidden Radiation

The most famous piece of spectroscopic history associated with gaseous nebulae and planetary nebulae is of course that of the bright green emission line at $\lambda 5007$. As is well known this line was a mystery for many years and indeed was thought to come from an unknown element which was named nebulium. The fact that most of the empty slots in the periodic table of the elements were rapidly filling up made it clear that there could be no mysterious nebulium. Quantum mechanics held the answer but it was I.S. Bowen again who in 1927 realised that this line together with one or two other very prominent lines were the result of forbidden transitions in doubly ionised oxygen, i.e. OIII (the standard notation for any forbidden transition is to put square brackets around the symbol for the element or ion in which it takes place; in this case we would put [OIII] $\lambda 5007$).

Neutral oxygen has two electrons in the 1s ($n = 1$, $l = 0$) level, two electrons in the 2s level and four in the 2p ($n = 2$, $l = 1$) level. In doubly ionised oxygen two of the 2p electrons have been removed and in the high-energy radiation field of say a planetary nebula this is caused by photo ionisation. OIII thus effectively becomes a two-electron atom with the 1s and 2s electrons bound tightly to the nucleus. As we saw towards the end of Chapter 3, a two-electron atom can produce spectral line series which are either single (singlets) or triple (triplets); furthermore, the lowest energy state for the 2p electrons is the triplet 2p level because with both electrons optically active, there is less shielding of the nucleus. Remember also that electron transitions which 'jump across' from a singlet level to a triplet level are forbidden by quantum mechanics. The reason for this is that with both electrons in singlet levels, one of the electrons is spin up and the other spin down; jumping across to a triplet level may involve a 'spin flip' whereupon both electrons end up with the same spin and this violates one of the selections rules. In particular, a transition from the 2p singlet level to the 2p triplet level would in fact be doubly forbidden because not only is there a potential spin flip but also the l quantum number would not change and as we recall, for a permitted transition this must change by 1. However as we know, the quantum mechanical selection rules are not hard and fast; they do get broken and the conditions within nebulae are just right for this to happen.

The 'Shelf Life' of an Electron

An electron absorbs a photon and jumps to a higher energy level; what then? In this case, what goes up definitely comes down and down it does come in a very short time; about 10^{-8} s. This is the typical time which an electron spends in an excited state before undergoing a permitted transition to a lower state. This 'electron shelf life' comes from some equations worked out by none other than Albert Einstein and which involve quantities which are called the *Einstein probability coefficients*. The derived times do vary but 10^{-8} s is very typical provided that the electron can leave the excited level by a permitted transition. A level from which there is no escape other than by forbidden transitions can hold an electron for much longer; say up to several seconds or even several tens of seconds. This is an eternity in the subatomic world and such 'long life' levels are called metastable. The 2p singlet level in OIII is a metastable level; the only way down is to the 2p triplet level and as we have seen this is a doubly forbidden transition. It does happen in nebulae though so we need to investigate further. There's also another 'agent' at work here.

'Rogue Electrons'

In any ionised gas, be it a planetary nebula or a hot stellar atmosphere, there will be free electrons whizzing here and there. Sooner or later a free electron will encounter an atom and as we have seen it may get captured by the atom. However, this doesn't always happen; the free electron may just have a close encounter with the atom and lose some of its energy. This energy can excite an electron within the atom, raising it to a higher energy level. This process is called not surprisingly, *collisional excitation*. It can also do something else; an electron within the atom which is in an excited state can actually have its excitation energy 'robbed' by the free electron. The atomic electron drops to a lower level but there is no emission of a photon (the free electron has made off with this energy) and this process is called *collisional de-excitation*. In the relatively dense conditions of a stellar atmosphere or for that matter a laboratory experiment, collisional excitations are followed almost immediately by either collisional de-excitations or indeed by emission or further absorption of photons. These frequent collisional de-excitations are important if a previous collision has excited an electron to a metastable level. Well before there's any chance of a forbidden transition, collisional de-excitation takes place.

In a nebula though things are different because the gas is of very low density and though the radiation field is of high energy with plenty of ultraviolet photons, it too is of much lower intensity than that within a stellar atmosphere. Photon-ionisation followed by recombination is a pretty rare event for an individual atom but the nebula is big so there are plenty of atoms undergoing these processes at any given time. An OIII atom with an electron in the 2p singlet level could 'sit there' for quite a long time before a photon excited the electron to a higher energy level or indeed it suffered a collisional excitation from a passing electron. In time even its long shelf life will run out and down it will drop to the 2p triplet level in a forbidden transition. The 2p triplet level is of course three levels and so three forbidden transitions are possible; one of these produces the λ5007 line and another also previously mysterious line at λ4959. The third possible transition is virtually not seen because it involves the inner quantum number *j* (see Chapter 3) changing by 2 which would break yet another selection rule.

OIII is not the only source of forbidden lines in nebulae; they come from OI, OII as well as ionised nitrogen (NII) and ionised sulphur (SII), etc.

The Edge of a Nebula

Images of planetary nebulae suggest that they often have fairly sharp outer edges; gaseous nebulae on the other hand are often seen to have edges which are diffuse. If a nebula is relatively dense then the supply of high-energy photons will eventually become exhausted; this happens over a relatively short distance giving the nebula a sharp edge. The actual size of such a nebula is thus determined by the supply of high-energy radiation and the nebula is said to be *radiation bounded*. The alternative scenario would be for the gas in a nebula to be relatively thin; in this case the high-energy photons keep on going but they have less and less target atoms to ionise. The nebula simply fades away at the edges and is said to be *matter bounded* or sometimes *density bounded*.

Summary

- Emission lines result from downward electron transitions.
- The high-energy radiation field within a planetary nebula or a gaseous nebula contains large numbers of ultraviolet photons.
- Ultraviolet photons ionise atoms in a process known as photoionisation.

- Recapture of an electron by an ionised atom is called recombination.
- Recombination usually takes place on a higher energy level which is then followed by a cascade of downward transitions resulting in the emission of optical wavelength lines.
- The overall result of photoionisation recombination is to degrade the ultraviolet photons into lower energy optical photons.
- Electrons can be excited to higher energy levels by collisional excitation caused by encounters with free electrons.
- Electrons can also suffer collisional de-excitation which results in no emission of photons.
- Excitation processes are rare for any individual atom but large numbers of these processes are happening at any one time because of the large size of nebulae.
- If an excitation process puts an electron into a metastable level a forbidden transition is more likely to take place before any further excitation or de-excitation occurs.

Glowing Vortices— Accretion Disks

Accretion disks are very familiar in astronomy these days; they are believed to exist around black holes, both in binary systems and at the centre of active galactic nuclei. They are also believed to form within many interacting binary stars; particularly cataclysmic variables or CVs. What's perhaps not quite so well known is that the hot gases which make up an accretion disk can produce emission lines of hydrogen and what's more these emission lines can have very striking profile shapes. They are in fact often double-peaked with what's called a central reversal in between; they also have broad extensive wings.

This chapter aims to do two things; firstly to give you an insight into astrophysical modelling. By modelling the double-peaked line profiles from accretion disks, astronomers can learn a great deal about cataclysmic variables. Secondly, accretion disks are an excellent example of how the large-scale motion of material can have a dramatic effect on line profile shapes.

Astrophysical Modelling

Most professional astronomers at some time or other have a need to develop some sort of model to explain what they see in their observations. The model may be that of a stellar atmosphere or perhaps a model to explain the behaviour of bipolar jets. The end result of developing a model will very likely be a computer program which will enable certain quantities to be varied so that the model can better match the observations; the matching procedure is actually called *fitting* and astronomers will often talk about producing a model *fit* to the observations. The quantities which are varied or maybe even just 'tweaked' are called *model parameters* and the set of all such parameters is referred to as the model's *parameter space*. A classic model program which is freely available to all via the Internet is the Wilson–DeVinney program for modelling the light curves of eclipsing binary stars.

Clearly any astrophysical model has to be firmly founded on the basic principles of physics and this means that modelling is often out of bounds for many amateur astronomers. However, if you feel you are up to it, there's no law which says you're not allowed to do this sort of thing. Indeed if you come across something during the course of your spectroscopic observations which is unusual and puzzling, you will inevitably be curious to know what's going on. Be warned though, once you have the beginnings of an idea it's often the easiest thing in the world to sit down and get carried away (I know, I've done it myself). You start to build into your model all kinds of things which will make the model more realistic; for example if your model involves the out flowing wind from a red giant star in a binary system, you say to yourself 'ah yes! The flow pattern of the red giant wind will swirl round due to the orbital motion of the binary'. Then you try to model the effect of this swirling motion on the profile of absorption lines which form as a result of a hot companion star shining through the wind. Unless you're very clever (and I mean the genius premier league here) you'll immediately be in serious guano as they say!

The rule with any sort of astrophysical modelling is to start very simple; our red giant wind model for example would start with material which was flowing from the star in straight lines. In this simple but not so realistic scenario it's much easier for example to work out the density and radial velocity of material in different regions of the wind, which in turn makes it easier to work out how much of the light from the hot star gets absorbed at a given wavelength. Once you've well and truly cracked the simple version of the model then you can start to add another level of complexity. Maybe though, doing this kind of modelling isn't really your thing; even so, if at any time you manage to secure some interesting spectra which you think might be of value to the professionals, it's always interesting to have some idea of the kinds of things that they might do with your precious data.

The spectra of cataclysmic variables contain emission lines of hydrogen; i.e. the Balmer lines which when observed at high dispersion are often seen to have very distinctive profiles which are double-peaked with a central reversal in between and broad well-developed wings as shown in Fig. 9.1. Their striking appearance virtually demands that they be modelled and the consensus of opinion among professional astronomers is that the source of these emission lines is the accretion disk which surrounds the white dwarf in these binary systems. So let's see if we can build a simple model to represent the emission line profiles which are observed, based on the assumption that they come from an accretion disk.

Another reason for doing this is to show that besides the main spectral line broadening mechanisms which we met in Chapter 5 it's also possible for line profiles shapes to be profoundly affected by large-scale motion of material and the revolving hot gas which makes up an accretion disk is a good example.

Anatomy of an Accretion Disk

Let's start simple; our accretion disk is a flat revolving disk of hot gas which as with most other things contains plenty of hydrogen. The disk is revolving around a central

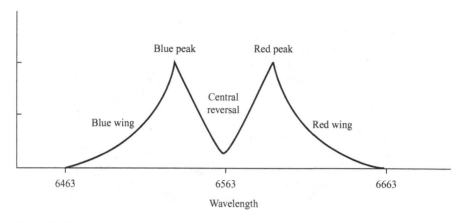

Figure 9.1. A stylised double-peaked emission line profile of the kind which might be emitted by the accretion disk surrounding the white dwarf in a cataclysmic binary.

star which might for example be a white dwarf. We could of course adopt a different type of star; say a main sequence star or even a neutron star; however as we shall see the model which we are going to develop can itself give us information on the type of central star (this is one of the main reasons for doing this kind of modelling) so at this stage it doesn't matter. The emission lines themselves, i.e. the Balmer lines, are of course recombination lines and so we need to get the hydrogen in the disk ionised. There are at least a couple of ways in which this might happen; firstly as the gas falls in the gravity well of the central star, loss of potential energy together with friction and viscosity effects within the gas could raise its temperature sufficiently to ionise the hydrogen. A second possibility is that ultraviolet radiation from the fast moving material in the inner regions of the disk could photoionise hydrogen in the outer regions just as hydrogen in a planetary nebulae gets ionised by ultraviolet radiation from the central star.

Again in the spirit of keeping things simple, we'll assume that the orbits followed by the atoms within the disk are circular. We also assume that even though on a small scale there will be encounters between atoms which will give rise to the above mentioned friction and viscosity, overall the motion of an individual atom is controlled only by the gravitational field of the central star. This in effect makes the atoms orbit just like tiny planets and as the motions of our neighbourhood planets obey Kepler's law of planetary motion, we say that the orbital motion of the disk material is *Keplerian*; we have a Keplerian disk.

Building the Model

We saw in Chapter 6 that rotation of stars can cause their absorption line profiles to be broadened due to the Doppler effect. For a revolving accretion disk, clearly the same thing is going to happen but the difference of course is that a star consists of a more or less spherical surface which is rotating whereas here we have a flat disk. Different

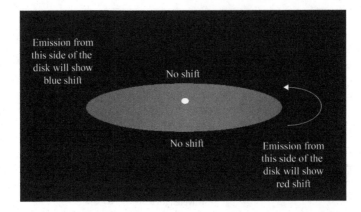

Figure 9.2. A revolving accretion disk as 'seen' by a distant observer; emission from one side of the disk is blue shifted while that from the other side is red shifted. Some material on the near and far sides of the disk is moving across the observer's line of sight and thus shows no Doppler shift.

geometry will cause the broadened line profile to be very different from that for a rotating star. However, just as a star seen 'pole on' will show no line broadening due to rotation, the same will be true of an accretion disk which is seen' face on'; all the material in the disk will be moving across our line of sight.

For an accretion disk which is presented more edge on, clearly one side of the disk will be moving towards us while the other side moves away and some of the material on the near and far sides of the disk will be moving across our line of sight as shown in Fig. 9.2. So the first thing to say is that emission from one side of the disk will be red shifted while that from the other side is blue shifted and in addition there will be some emission which doesn't show any shift. Emission lines which are produced by the revolving disk material will of course be subject to thermal broadening and there will almost certainly be some turbulence broadening too but for the moment we'll ignore these; they can as it turns out quite easily be incorporated into the model at a later stage. So the Doppler shifts produced by the various parts of the disk will be the main line broadening mechanism. If you think about it, each point on the observed emission line profile will be the combined result of all the emission which comes from the disk and which also has the same radial velocity. So the next thing we need to think about is the distribution of radial velocity across the surface of the disk.

Zero Radial Velocity—The Line Profile Centre

By convention, a disk which is seen face on has a tilt or an *inclination* of 0° while an exactly edge on disk has an inclination of 90° and clearly the rotational line broadening will be greatest for an edge on disk. So let's imagine a disk with an inclination of 90°; another reason for doing this is that in this case the orbital velocity at any point on the disk actually equals the radial velocity and remember that radial velocities translate directly into wavelength shifts. Firstly, if we bisect the disk with a line pointing in the direction of the observer as shown in Fig. 9.3, all material crossing this line also crosses the observer's line of sight. This material has zero radial velocity and so emission from here is not Doppler shifted; it marks the centre of the line profile.

The Wing Limits

If we now bisect the disk with another line at right angles to the first line, then clearly emission from material crossing this line will be seen to be either blue or red shifted depending on which side of the disk it is. Note here though that the speed and hence the radial velocity of orbiting material varies as we move along this line; stuff at the outer edge of the disk will be moving more slowly and hence have a smaller Doppler shift than material at the disk's inner edge. In fact, the two points which are on this line and at the disk's inner edge will have the highest radial velocity of all and will thus mark the wing limits of the line profile. This maximum radial velocity will be determined by the radius of the disk's inner edge and also the mass of the central star; the higher the star's mass and the smaller the disk inner radius, the higher the orbital velocity and hence the higher the radial velocity.

There is yet again a simple plug in formula which we can use to work out what the maximum orbital velocity for accretion disk material might be if we're talking about a cataclysmic variable with a white dwarf as the accreting star. This formula comes straight from Kepler's laws of planetary motion; it says that the orbital velocity which we can call 'v' is simply given by

$$v = \sqrt{\frac{G \times M}{R}} \qquad (9.1)$$

Here G is the universal constant of gravitation and it equals 6.673×10^{-11}. M is the mass of the accreting star and R is the distance from the star. Let's stick with our CV binary; we can take the mass of a white dwarf to be equal to the mass of the Sun which is 1.99×10^{30} kg. A typical white dwarf has a radius of about one hundredth that of the Sun; i.e. about 6.96×10^6 m. Let's now assume that the inner edge of the accretion disk is just about on the surface of the white dwarf so the inner disk radius equals the white dwarf radius. Try plugging these numbers into Eq. (9.1); (don't forget about the

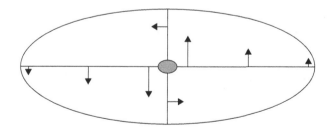

Figure 9.3. Orbiting disk material which crosses the horizontal line has a range of radial velocities as indicated by the size of the arrows; material at the disk's inner edge has the highest radial velocity.

square root) the answer comes out at about 4400 km/s. At this velocity it's probably still okay to use the simple Doppler shift formula and if you do this for the Hα line you'll see that the wing limits of the line profile will be shifted by about 100 Å. Of course the really neat thing to do is to determine the wavelengths of the wing limits from your spectrum (see the method for doing this in Chapter 5); these then translate directly into the radial velocity of material orbiting at the inner edge of the disk.

What you don't get from Eq. (9.1) of course is the orbital velocity unless you know for certain that the accretion disk is being observed exactly edge on; and it's a pretty fair bet that you don't know this. The radial velocity 'v' is in fact equal to the orbital velocity multiplied by the sine (remember sines, cosines and tangents from high school trigonometry?) of the disk's inclination angle. It's clear though that a more edge on disk will mean higher radial velocities and hence broader wings; the calculation which we've just done probably represents an upper limit to the kind of radial velocities which would be encountered in cataclysmic binaries which involve a white dwarf. This upper limit results from an edge on disk, i.e. the inclination angle equals 90° and sin(90°) equals 1. So for example, if the wing limits of your profile represent wavelength shifts of only about 50 Å then this means that the radial velocity is only about half that which would be expected for an edge on disk. This means that we would have to multiply the orbital velocity by a half to get the observed radial velocity. This means that the sine of the inclination angle equals 0.5; sin(30°) equals 0.5 and so this gives us at least an estimation of the orbital tilt of this binary system of about 30°—this is real amateur astrophysics! Finally, as we'll see below there's also another check which we can do to tell us whether the inclination angle of a binary system is low or high.

The Emission Line Peaks

The next thing to do is to start at a point on the horizontal line in Fig. 9.3, which is at the outer edge of the disk and mark out points on the disk which have the same radial velocity as this point. As we move away from the line we have to do two things to keep the radial velocity constant; firstly we need to move either clockwise or anticlockwise around the disk slightly. This by itself would lower the radial velocity but if at the same time we move slightly inwards towards the centre, the slight increase in orbital velocity will in turn slightly increase the radial velocity. This combination keeps the radial velocity constant and enables us to trace out a curve of constant radial velocity on the disk surface. Repeating this procedure systematically for successive points along the line enables us to map out a whole series of constant radial velocity curves. This striking pattern is shown in Fig. 9.4 and clearly resembles the pattern of field lines produced around a bar magnet; indeed this pattern is sometimes referred to as a dipole field pattern.

Emission from all the way along one of these constant radial velocity curves will combine to produce one point on our emission line profile. Looking at the dipole field pattern we can see that the physically longest curve is the one which sweeps inwards from the outer edge of the disk. So we'd expect emission from this curve to coincide with peak emission in the line profile and two such curves on opposite sides of the disk give us the red and blue emission peaks in the profile. So in our model, the emission peaks coincide with the radial velocity of material which is orbiting at the disk's outer edge and the wavelength shift of each peak can be used to determine this

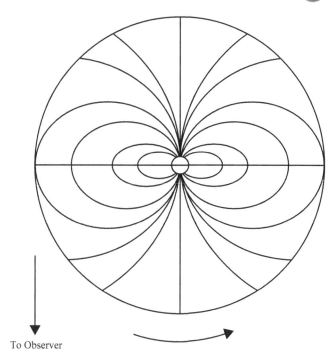

Figure 9.4.
The striking 'dipole field' pattern of curves of constant radial velocity on the surface of an inclined rotating accretion disk. Notice that the physically longest curves are those which sweep inwards from the outer edge of the disk.

To Observer

radial velocity. If we've already managed to estimate the inclination of our CV binary from the wing limits, we can now use this to determine the orbital velocity of material on the disk's outer edge. Finally, by rearranging Eq. (9.1) we can even estimate the radius of the accretion disk itself.

Building the Line Profile

Notice now that as we move from one of the peak emission curves towards lower radial velocity values (i.e. towards the vertical line in Fig. 9.4), what would be longer curves get cut off or truncated by the disk's outer edge so their length decreases as we move towards the zero radial velocity line. This results in lower emission values as we move towards the centre of the line profile and this gives us our central reversal. Finally, as we move towards the higher radial velocity region (by moving inwards along the horizontal line in Fig. 9.4), the curves shrink rapidly making the line profile fall away towards the wings. Fig. 9.5 shows the disk radial velocity pattern again with several areas shaded. Similarly, shaded areas on the accompanying stylised line profile show which parts of the disk contribute to which parts of the line profile.

If the disk is seen face on then obviously the dipole field pattern disappears because all material in the disk has zero radial velocity. As soon as we tilt the disk the dipole pattern appears but because the disk inclination is only slight, the range of radial velocities in the dipole pattern is small. The line profile would be very narrow and though technically double peaked, any central reversal would be extremely slight. As the inclination of the disk increases the line broadens as the range of radial velocities

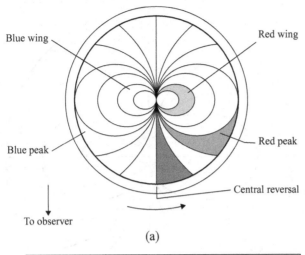

(a)

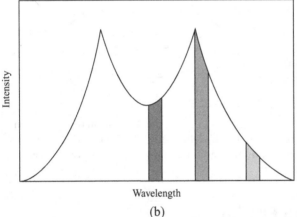

(b)

Figure 9.5. 'a' and 'b' show which parts of the accretion disk produce which parts in the line profile, as indicated by the shaded areas.

increases. The difference in length between the disk outer edge radial velocity curve and that for the zero radial velocity line also increases; this translates into a bigger difference in the total emission from these curves and so the central reversal deepens until at an inclination of 90° the central reversal is at its deepest and the line profile has the broadest wings as shown in Fig. 9.6. Indeed the central reversal depth can be used as a kind of check on the inclination of the binary system; a deep central reversal strongly suggests that the binary is being observed more edge on. This is finally verified by the peak-to-peak separation which is wider for a more edge on system.

Enhancing the Line Profile Wings

So far we have assumed that emission values are the same all over the disk or to be more precise, the emission from every square metre of the disk's surface is the same. This means that emission vales are constant along any given constant radial

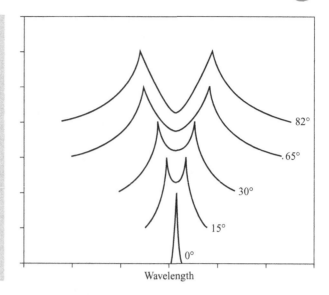

Figure 9.6. A series of model line profiles computed for a range of inclinations; notice how the overall line width increases together with the peak-to-peak separation, as well as the central reversal depth as the inclination increases.

Wavelength

velocity curve. Material orbiting within the disk gradually spirals inward; the exact details of how this happens are still not fully understood but the process is relatively gradual and results in a build up of material towards the inner regions of the disk. More material means more emission per square metre and so to produce a more realistic model emission values should rise as we move along a constant radial velocity curve towards the centre. The overall result is that emission from the inner regions of the disk will be enhanced. Again, looking at the dipole field pattern, it's clear that emission in the line profile wings comes predominantly from the inner regions of the disk and so more material here means enhanced wings in the model profile.

To calculate and plot model emission lines profiles like these obviously involves a lot of work but just by using simple reasoning, we've managed to explain why emission line profiles from accretion disks are double peaked and the wavelength separation of the peaks depends on the radial velocity of material orbiting at the disk's outer edge. The limits of the profile wings are determined by the radial velocity of material orbiting at the disk's inner edge and the overall width of the line profile together with the depth of the central reversal depends on the inclination of the disk to our line of sight. We finally showed that as material 'piles up' in the inner regions of the disk, this will produce enhanced emission in the profile wings.

A Better Model

The model which we've described was first developed by J. Smak in the late 1960s. The central reversals in Smak's model line profiles were relatively shallow and 'U' shaped even when the binary system was assumed to be nearly edge on. The Hα line profiles of cataclysmic variables were often seen to have deeper more 'V' shaped central reversals and in the mid-1980s the astronomers Keith Horne and Tom Marsh developed a better and more realistic model to explain this. The model line profiles shown in Fig. 9.6 were actually computed using the Horne and Marsh model.

One simple feature of Smak's model is that the accretion disk gas is of low density, notwithstanding the higher densities towards the centre. The significance of this is that emission line photons which are produced within the disk can basically escape unhindered; in astrophysics terminology we say that the accretion disk is *optically thin*. Balmer emission lines produced within an optically thin gas would show a very predictable Balmer decrement (the ratios of the intensities of the lines in the Balmer series as explained in Chapter 8). Horne and Marsh realised that in the case of CV spectra the line intensities decreased less rapidly (this is sometimes called a 'flat' Balmer decrement) and the explanation for this is that the accretion disk material must be relatively dense. Denser material means that emission line photons get trapped within what is now an *optically thick* disk. Line photons do get out however, otherwise we wouldn't see the emission lines at all—so let's see how they manage to escape.

Besides assuming that the disk material is relatively dense, we're also going to give the disk a finite thickness, though this thickness is still assumed to be small compared to the diameter of the disk. This means that an emission line photon which is produced within the disk will have to 'dodge' potentially absorbing atoms which will get in its way, if it's going to escape. In this sense the disk is behaving very much like the photosphere of a star; some photons will make it, while others will get absorbed or more likely scattered and sent off in a different direction only to be scattered again and again. There is however an additional effect which is the key to the Horne and Marsh model. Imagine a cross section of part of the disk as shown in Fig. 9.7; a λ6563 photon which travels from within the disk to the disk surface will generally pass through a small but finite range of distances from the central star. This means that along the photon's route the atoms are orbiting at increasingly different velocities to that of the atom which emitted the photon.

For these atoms to stand a chance of absorbing the photon, their radial velocities must be zero with respect to the atom which emitted the photon. This means that they 'see' the photon as λ6563, otherwise they 'see' the photon with a Doppler shifted wavelength and ignore it; the photon escapes.

Have a look at Fig. 9.8; if the photon were travelling directly away from the central star, i.e. in an outward radial direction, then while atoms along its route would have slightly different orbital velocities they would all have zero radial velocity relative to the atom which emitted the photon. These atoms 'see' the photon at λ6563 and are only too happy to absorb it. So a photon doesn't stand much chance of getting out along this direction. Now take a photon which travels through the disk in a direction at right angles to the radial direction; along this line all the atoms including the one which emitted the photon have the same orbital velocity and the photon itself is travelling in the same direction as the atoms. These atoms again 'see' the photon as λ6563 and the photon very likely gets absorbed. Finally, take a photon which travels in a non-radial direction, i.e. not directly away from the central star. Once again there are atoms with different orbital velocities along the way but these atoms also now have different radial velocities relative to the atom which emitted the photon; they don't see the photon as λ6563 so they 'let it go' and the photon escapes.

The overall effect is that photons which travel through the disk in a radial direction, together with those which travel at right angles to this direction stand less chance of

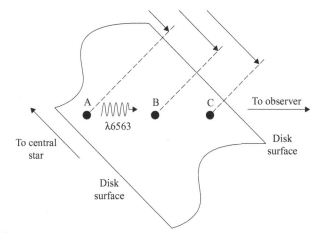

Figure 9.7. Atom 'A' emits a λ6563 photon which travels towards the surface of the disk. Atoms 'B' and 'C' which lie along its path have different orbital velocities to atom 'A' because they are at different distances from the central star. In order for either 'B' or 'C' to stand a chance of absorbing the photon, they must have zero radial velocity with respect to atom 'A'.

escaping than those which travel in a direction somewhere in between. In fact, the most favoured route of escape is for the photon to travel through the disk at 45° to the radial direction because along this line the photon encounters the greatest range of radial velocities so the atoms leave it alone. This range of radial velocities across the thickness of the disk is called by Horne and Marsh the *shear velocity*. Its over-all effect on the line profile is to deepen the central reversal into a 'V' shape which matches real observed line profiles much more closely. Horne and Marsh added an extra level of complexity but also an extra level of realism to the Smak model; in some ways it seems merely like an over fussy detail, but here the details are every-thing.

Thinking Up an Even Better Model

So far we've seen how to develop a pretty good model for producing simulated line profiles which can be used to match the real thing. The model still has its limitations though; Horne and Marsh admit that their model breaks down for disk inclinations of more than about 87° but an even more obvious limitation is the fact that as it stands, the model can only produce profiles with emission peaks of equal height because the distribution of radial velocity across the surface of the disk is totally symmetrical. It's generally reckoned that the accretion stream—the stream of gas issuing from the mass-losing star in the binary system, causes a 'hot spot' as it hits the outer rim of

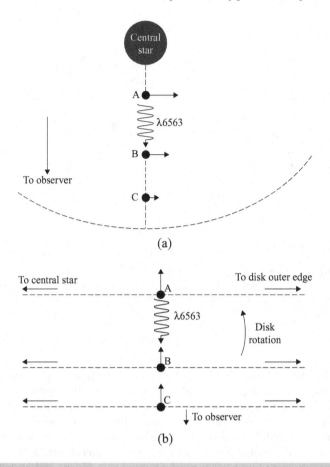

Figure 9.8. Atom A lies nearest to the rear surface of the disk; atom C is closest to the front surface and B and C lie along the path of the photon emitted by atom A. In case a; B and C have slightly different orbital velocities to A as indicated by the different lengths of the velocity arrows but they are moving across A's line of sight; there is zero radial velocity between the atoms, so B and C 'see' the photon as λ6563 and are more likely to absorb it. In case b; we're seeing a very small area of the disk and over this small area atoms B and C have the same orbital velocities as A; again there is zero radial velocity between the atoms and the photon very likely gets absorbed. In case c; the photon is emitted at an angle to the radial line drawn from atom A to the central star. Atoms B and C now also have non-zero radial velocities relative to atom A as well as different orbital velocities. They 'see' the photon Doppler shifted—the photon escapes.

the disk. Some authors even argue that the stream initially passes straight through the disk and out through the far rim on a highly elliptical orbit, before re-entering and causing yet a further hot spot. These hot spots are other sources of emission whose apparent location on the disk will vary as the binary rotates; this could clearly cause

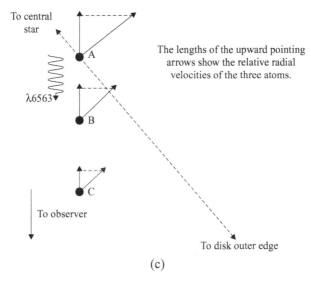

The lengths of the upward pointing arrows show the relative radial velocities of the three atoms.

(c)

Figure 9.8. (*Continued*)

asymmetry in the emission peak heights and also affect the central reversal depth. We could obviously go on for ever, but the further we go the harder it gets so I guess now is the time to stop. Even if you're not sure about the physics (ask a friend with a physics degree) and don't like the math (I'm sure he could do this too) thinking up ever more realistic models is great fun—for professional astronomers too.

One final word; just because you've thought up and developed a model which seems to give great fits to your observations doesn't mean to say that it's the correct model. If someone else out there has an alternative model to explain the same thing, the chances are they'll want to shoot your model down in flames, so be prepared to have to defend it.

Summary

- Astronomers develop 'models' to try to explain their observations.
- Models should always start very simply with only a gradual addition of more complexity or realism.
- Emission lines from accretion disks are double-peaked with broad wings and a central reversal.
- The double-peaked profile shape results from the distribution of radial velocity across the surface of the disk.
- A simple model to explain disk line profiles allows photons to escape freely from the disk—the disk is said to be optically thin.

- A more realistic model which gives a better fit to the observations assumes an optically thick disk which also has a finite physical thickness.

- Photons can more readily escape along directions where there is a greater range of radial velocities or shear velocity across the thickness of the disk.

- Accretion disk line profiles are a good example of how spectral line profiles can be affected by the large-scale motion of material.

The P Cygni Profile and Friends

If there's one spectral line profile which most amateur astronomers have heard of it's the P Cygni profile. P Cygni itself is a class B star lying in the plain of the northern Milky Way close to the second magnitude star Gamma Cygni. P Cygni profiles are all to do with stellar winds and they come as we shall see in various interesting guises.

The Classic P Cygni Profile

The classic P Cygni line profile is shown in Fig. 10.1; it consists of a broad intense emission line with a less intense and narrower absorption line displaced to the blue side of the emission line. This famous line profile shape is caused by large-scale motion of material (i.e. hot gas) but this time the moving material is an outflowing wind from the star.

In some ways we have a situation here not too dissimilar to that of a planetary nebula; ultraviolet radiation from the hot star's photosphere ionises the surrounding wind material. Recombination then causes this material to produce emission lines. There is a big difference here though; whereas the rate of outflow of material in a planetary nebula is pretty low; only perhaps a few kilometres per second, the outflow rate in a hot star's wind can easily reach several hundred kilometres per second and in the case of Wolf-Rayet stars as much as 1000 km/s. Just as rotational motion is the main cause of line broadening in accretion disks, with outflowing stellar winds it is the motion of the wind material itself which broadens the line profile. A faster wind will produce broader emission lines and in the case of Wolf-Rayet stars, emission lines become broad emission bands. Because of the high temperatures involved with these hot stars, spectral lines due to very highly ionised elements are produced and many

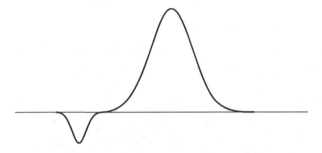

Figure 10.1. The classic P Cygni profile consists of a broad intense emission line together with a less intense and somewhat narrower absorption line displaced to the blue side of the emission line.

of these lines lie in the ultraviolet part of the spectrum. However, lines due to neutral and ionised helium (HeI and HeII) and sometimes the Hα line can be seen as P Cygni profiles in the optical region.

Wind Outflow Geometry

Interpreting the P Cygni profile in terms of what's happening in the neighbourhood of a star which is producing these kinds of lines is really quite easy; though there is one point which perhaps needs a little clarification. Have a look at Fig. 10.2. Here

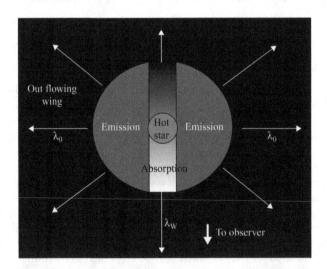

Figure 10.2. This is how the P Cygni profile is produced. The out flowing wind is photo ionised by the hot star and produces emission lines which are broadened primarily by the flow of wind material. Absorption in the wind region between the star and the observer produces the blue-shifted absorption line.

we have a hot star with a fast outflowing wind which is spherically symmetric, i.e. the same in all directions. The atoms which make up this wind are photoionised by high-energy radiation from the stars and so the wind is literally glowing with emission line radiation. Seen by a distant observer there is clearly a region of this wind which is hidden behind the star and so our observer sees nothing of this region which would produce the emission with the largest red shift. The regions of the wind which are seen on either side of the star are responsible for the emission line profile; they incorporate a range of Doppler shifts both blue from the near or approaching side of the wind and red from the rear receding region. There is also material which is moving across our line of sight and the result is an emission line which is broadened by these Doppler shifts. The material which is crossing our line of sight corresponds to the wavelength of the center of the emission line.

Now let's think about the wind material which is coming straight towards us and lies directly between us and the star. This stuff is responsible for the absorption line and because this material is aimed straight at us, it has the largest blue shift. Hang on though! Isn't this material supposed to be producing emission line radiation too, just like the other parts of the wind? How then can it produce an absorption line? First, think about the wind regions on either side of the star; this material absorbs photons from the star and these photons are coming from one direction only—namely from the star itself. When photons are emitted by this material, they come out in all directions so there is a net loss of photons going straight out from the star; put another way, without the wind material there, all photons would head out straight away from the star but with the wind material present, many of these are lost, recycled and sent out in all directions. So, many of the photons which were coming straight towards us from the star have been sent off in other directions by the wind material and this net loss shows itself as a (blue shifted) absorption line.

A useful bit of information which can be obtained directly from a spectrum is the velocity of the stellar wind; this comes simply from the difference in wavelength between the center of the emission line and that of the absorption line. Simply plug this wavelength difference into the simple Doppler effect formula to get the wind velocity. Even at Wolf-Rayet star wind speeds the simple Doppler formula should be quite sufficient.

P Cygni Profiles from Cool Stars

P Cygni profiles are most famous for their connection with the hottest stars; the combination of emission lines with a fast outflowing stellar wind gives us the classic profile shape. However, the defining feature of a P Cygni profile is not so much the emission line but the blue-shifted absorption line, because this alone is the telltale signature of the wind. Cooler stars have winds too; our own sun has its solar wind and even red giants have outflowing winds. In the case of a red giant the wind is a slow one—maybe 10 or at most 20 km/s but the wind material can be relatively dense and so we'd expect there to be some kind of absorption caused by this wind. Red giants don't have emission lines as a permanent feature of their spectra but they do have plenty of absorption lines in addition to the molecular bands. Extra absorption is produced by the wind material and this can show itself as an additional absorption

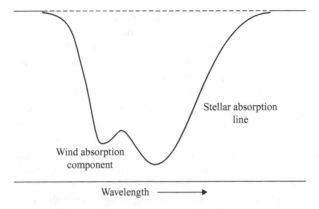

Figure 10.3. A stylised P Cygni profile in the spectrum of a red giant; there is no emission, but a stellar absorption line may include a blue-shifted component caused by the slow out flowing wind of the star.

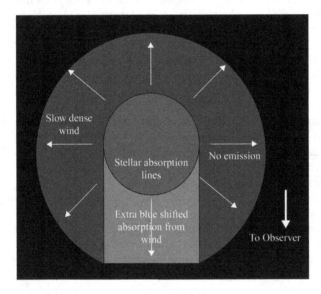

Figure 10.4. In some ways the scenario for producing P Cygni profiles in cool giant spectra is similar to that for hot stars, except that the out flowing wind is much slower and produces no emission lines.

component to a stellar absorption line which is shifted to the blue side of the line center as shown on Figs. 10.3 and 10.4.

In some ways the setup here is similar to that for a hot star, i.e. outflowing wind, absorption by wind material, etc. The big difference is obviously that the environment here is too cool for the wind to produce emission lines; nonetheless this blue-shifted absorption component does classify the line as a type of P Cygni profile.

A P Cygni Profile Mystery—Symbiotic Stars

Here's yet another type of P Cygni profile and this time its exact cause is still subject to much debate among astronomers. Fig. 10.5 shows the Hα line profile of the symbiotic star AX Persei; here we have an emission line with broad wings and what certainly

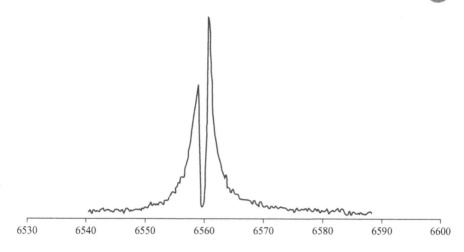

6530 6540 6550 6560 6570 6580 6590 6600

Figure 10.5. The Hα line profile of the symbiotic star AX Persei showing a slightly blue-shifted absorption component.

looks like a deep absorption component just to the blue side of the line center. This absorption component suggests a dense but slow outflowing wind, which from what we saw above seems to be the hallmark of a red giant. The emission line itself clearly tells us that there's something hot within this system and the extended wings suggest fast moving material; however there isn't an additional absorption component with a larger blue shift which would suggest absorption in an outflowing fast wind. So here we have a puzzle!

First a word or two about symbiotic stars which might give us one or two clues; they are variable stars of course but they're something of a 'rag bag' bunch in the sense that there is no standard model for what a symbiotic star actually is. In fact, they are in a sense almost totally defined by the nature of their spectra which generally consist of three ingredients. Firstly, the spectrum of a cool giant is present often of class M with all the usual absorption lines together with bands due to molecules. Secondly, there are virtually always emission lines, not just the Balmer lines but lines which involve much higher ionisation potentials. Finally, there is a continuum in the blue region of the visible spectrum; this kind of feature is of course not seen in the spectra of ordinary red giants. The blue continuum is usually ascribed to what researchers call a 'hot source' within the system and it's generally reckoned that this hot source somehow produces the emission lines. The bottom line to this is that symbiotic stars are regarded by most as binary stars with a red giant, a hot star of maybe class O or B and surrounding gas which gets ionised by the hot star.

Another possibly important clue is that while the absorption component is almost always just to the blue side of the emission line center, this is not always the case; Fig. 10.6 shows the Hα line for the star BX Monocerotis, where we have the opposite situation with the absorption component just to the red side of the line center. In the world of P Cygni profiles, this is usually taken to be the result of in falling material rather than an outflowing wind. Imagine reversing the direction of the wind flow arrows in Fig. 10.2; the emission line profile would remain unchanged but the absorption line would now shift over to its red side.

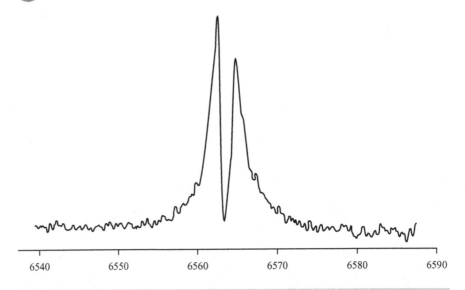

Figure 10.6. The Hα line profile of the symbiotic star BX Monocerotis showing a slightly red-shifted absorption component.

The small Doppler shift for the absorption component certainly suggests that we're reasonably safe in assuming that this is indeed caused by the red giant wind. However, one thing that we do have to bear in mind is that here we're dealing with a binary system and not just a single star with an outflowing wind. It may well be that in many instances we see the emission line (produced in some way by the hot star) shining through and suffering absorption in the outflowing wind of the giant. Occasionally though we appear to be witnessing the emission line shining through red giant wind material which is perhaps falling towards the hot star. This in itself raises the question as to the exact location of the source of the emission lines; are they produced in close proximity to the hot star or are they a less localised phenomenon?

Let's take the close proximity idea first; this suggests of course that the hot star can accrete material from the red giant wind. It has even been suggested that some symbiotic stars may contain an accretion disk just like cataclysmic binaries. Unlike their smaller cousins, however, symbiotic accretion disks are likely to be the result of accretion from the wind rather than by Roche lobe overflow which produces accretion disks in CVs. Material in whatever form in close proximity would be an obvious source of emission lines because it would be photoionised by the high-energy radiation from the hot star. Material close to the hot star would also be moving rapidly and this would fit the bill for explaining the broad wings in the line profile. An emission line shining through in falling or outflowing wind material would determine the position of the absorption component in the spectrum by why we predominantly see a blue-shifted component but sometimes a red-shifted one is still open to debate. Another feature of these symbiotic P Cygni profiles is that they do vary in prominence and appearance. As the professionals say 'more observations are needed' and amateur spectroscopists who can take high-resolution spectra could do no better than monitor these strange stars.

We're not quite finished with symbiotic stars; many if not more of them show emission lines which are fairly narrow and show no trace of any absorption component. In these cases it's likely that the hot star, rather than accreting red giant wind material is simply photoionising a large part of the wind itself turning it into some kind of small dense planetary nebula. Indeed symbiotic spectra can also contain forbidden lines due to doubly ionised oxygen just like planetary nebulae—truly remarkable stars!

In conclusion, P Cygni profiles can show up in probably any situation which involves absorption in an extensive area of material surrounding a stellar system.

Summary

- The classic P Cygni profile consists of a broad intense emission line with a blue-shifted absorption component.
- P Cygni profiles are another example of how line profile shapes are determined by the large-scale flow of material in and around stellar systems.
- The defining feature of a P Cygni profile is not an emission line but rather a Doppler shifted absorption component. It's possible to have P Cygni profiles without emission lines, e.g. in the spectra of red giants.
- Symbiotic stars often show P Cygni profiles in their spectra but their interpretation is still open to debate.

Spectral Magnetism—The Zeeman Effect

As kids we all played with magnets; these pieces of metal show more dramatically than anything just how a force can act across empty space. We probably all played with a compass too and perhaps like the young Albert Einstein were mystified as to how the needle kept pointing in the same direction no matter which way we turned the compass. Magnetic fields are produced by electric currents or flows of electric charge; the outer or optically active electrons in an atom constitute tiny electric currents which generate tiny magnetic fields. If a population of these atoms find themselves in an external magnetic field their tiny magnetic fields respond to produce a wonderful phenomenon called the Zeeman effect in honour of the Dutch physicist Pieter Zeeman.

How Strong Is a Magnetic Field?

A bar magnet has two poles; a north pole and a south pole and as we may remember from high school physics, two like poles will push each other apart, whereas two unlike poles will try to pull themselves together. If we try to push two north poles towards each other, the force trying to push them apart gets stronger as they get closer together. Clearly the strength of the magnetic field surrounding the magnet increases as we get nearer to the pole itself. The standard unit of magnetic field strength which is most commonly used in astronomy is called the 'gauss'; named after Carl Friederich Gauss. You may also come across another unit which is used widely by physicists called the 'tesla' named in honour of a Croatian-American electrical engineer called Nikola Tesla. One tesla is equal to 10,000 (10^4) gauss.

The best way to get a 'feel' for what might be unfamiliar units like these is to first of all take the Earth's magnetic field; this has a strength of about 0.3 G at the equator to about 0.6 G at the magnetic poles. The Sun's overall magnetic field strength is about 1 to 10 G but in sunspots, magnetic fields can reach strengths of 1500 to 3000 G (this is 0.15 to 0.3 T). Among the most powerful magnetic fields encountered in the universe are those of magnetic white dwarf stars which have the awesome strength of from 100 to 10^5 T (10^5 tesla is equivalent to 1 billion gauss!). Even these though pale by comparison with the fields of magnetic neutron stars or magnetars which can reach 10^{11} T.

More on Electrons in Atoms

Way back in Chapter 3 we learned that the l sublevels in an atom are themselves divided into sub-sublevels which are identified by the magnetic quantum number m_l. Remember that m_l has whole number values (including zero) running from $-l$ to $+l$; so for example the $l = 3$ level splits into seven m_l levels. However, most of the time these m_l levels are dormant and correspond to the same energy for a given value of l; now it's time to 'wake' these 'sleeping' levels by applying a magnetic field. When we do this the m_l levels separate out and have slightly different energies; $m_l = 2$ has slightly higher energy than $m_l = 1$ which in turn has slightly higher energy than $m_l = 0$ and so on.

So far, aside from the energy differences between levels, these quantum numbers have been little more than a kind of code or address system for telling us 'whereabouts' in an atom an electron actually is. The magnetic quantum number clearly has important physical significance for electrons in atoms which are subjected to an outside magnetic field. To understand this significance and the role it plays in the Zeeman effect, we need to look at a very interesting bit of physics.

Momentum

The idea that energy is something which can do work and move things around is very familiar; perhaps not quite so familiar is the nature of *momentum* even though the word itself is very common. Energy can be possessed by something which is motionless; this kind of energy is not surprisingly called potential energy. For example, an apple hanging on a tree has the potential to fall and on the way down it could hit a small lever on the other end of which was a ping-pong ball which could be thrown through the air by the action of the falling apple. There is also energy due to an object's motion and this is called kinetic energy. Momentum though is something which exists only because an object is in motion; there is no version of 'potential momentum'. Any moving object has an amount of *linear momentum* whose value is simply equal to the object's mass multiplied by its velocity. A massive object like a truck can have a large amount of linear momentum even if it's only moving slowly and a much less massive object like a bullet can also have a large amount of linear momentum because it travels so fast.

Angular Momentum

There is another form of momentum called *angular momentum*; this is possessed by bodies which are revolving like the Earth or moving in an orbit around another body—again like the Earth orbiting the Sun. I remember a physics exam question which I had to answer in high school which asked what was the total momentum possessed by a perfectly smooth pool ball rolling along a perfectly smooth pool table. The ball obviously has linear momentum because it's moving along the table and I would have to stick out my finger to stop it. If my finger were also perfectly smooth though, the ball even when stopped would still keep revolving—it has angular momentum. I'd need to apply a twisting action or a torque with non-smooth fingers to stop it revolving.

For an orbiting object the amount of angular momentum is again equal to the object's mass multiplied by its velocity but this time multiplied also by the radius of the orbit. There's also another very important feature of angular momentum; it has *direction* as well as an actual value and is what physicists call a *vector* quantity. How do we give a sense of direction to something which is say, moving around some kind of circular orbit? Clearly a bullet has a direction as it travels through the air and its linear momentum which is also a vector quantity, would have the same direction. For an object moving in a circular orbit, its direction is changing all the time; though we do know that in the course of its motion, its direction will change by 360°. If the orbit is stable the actual value of the object's angular momentum will stay fixed; so the only thing which remains to say about it is whether it is moving around its orbit in a clockwise or an anticlockwise manner. If we observe the object to be moving anticlockwise, then an arrow pointing from the centre of the orbit straight towards us can be used to indicate this. Conversely, an arrow pointing directly away from us can be used to show that the object is moving clockwise; these arrows are always drawn at right angles to the plane of the orbit and what's more we can use the length of the arrow to indicate the actual value of the object's angular momentum. This then is how the angular momentum of an object moving in an orbit is represented; indeed the arrow itself is often referred to as the *angular momentum vector*. In a similar way, the angular momentum of a revolving object such as a planet can be represented by an arrow of length equal to the object's rotational angular momentum and pointing from the centre of the object towards its 'north pole'. The north pole of a planet is

Figure 11.1. This is how the angular momentum vector is used to tell whether an object (like an electron) is orbiting clockwise or anticlockwise. The length of the arrow is used to fix the value of the angular momentum.

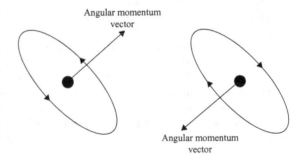

Angular momentum vector

Angular momentum vector

Value of electron's
angular momentum
depends on *l*

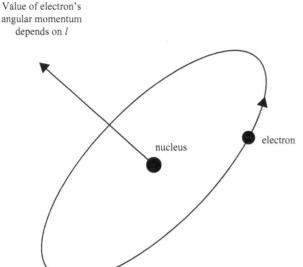

Figure 11.2. An electron's angular momentum vector is represented like this; the value of the angular momentum is determined by the angular momentum quantum number *l* and we can set the length of the arrow equal to *l*.

nucleus

electron

defined by the condition that when you look down on the pole, the planet is revolving anticlockwise.

An electron moving around the nucleus of an atom has angular momentum and indeed the *l* quantum number is called the angular momentum quantum number because it can be used by physicists to calculate the actual value of the electron's angular momentum. The electron's angular momentum vector would then be an arrow of length equal to this value and pointing from the atom's nucleus in the appropriate direction at right angles to the plane of the electron's orbit.

The Wonderful World of *x y z*

Representing an electron's angular momentum as an arrow pointing from the atomic nucleus in the relevant direction may seem a bit strange at first, but it provides us with a very important piece of information; it tells us about the electron orbit's orientation in space. In exactly the same way, the direction of the Earth's rotational angular momentum vector tells us that the plane of the equator makes an angle of about 23° to the plane of the orbit, i.e. the ecliptic. To be more specific, we can choose some direction in space and call it the *x* direction; the opposite direction would then be called the −*x* direction. Space is three-dimensional so a direction at right angles to the *x* direction can be called the *y* direction (with a corresponding −*y* direction) and finally the direction at right angles to both the *x* and *y* directions would be called the *z* direction. We can talk about distances measured along these three directions from some chosen starting point; what we have now is a *reference frame* or as it's sometimes called; a *coordinate frame*. The *x*, *y* and *z* directions are then referred to as the *x*, *y* and *z axes*.

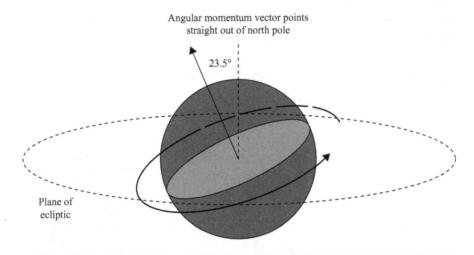

Angular momentum vector points
straight out of north pole

23.5°

Plane of
ecliptic

Figure 11.3. The direction of the Earth's rotational angular momentum vector tells in effect that the plane of the equator is tilted by 23.5° with respect to the plane of the ecliptic.

We could now arrange our coordinate frame so that our electron's angular momentum vector points straight up the z-axis; the electron's orbital plane is then the xy plane. This would make life relatively simple but in real spectroscopy we are of course dealing with vast populations of atoms and we can't expect all the angular momentum vectors to conveniently line up along the z-axis; any electron in an individual atom will have its angular momentum vector pointing in some arbitrary direction which makes a different angle with the three axes. However, once we've actually established our coordinate frame, the orientation of any electron's angular momentum vector can be specified in terms of these three directions. Again, being more specific, to get from the foot of our angular momentum vector (at the nucleus) to its tip we need to move in the correct direction by a distance that corresponds to the value of the electron's angular momentum. We can also do this by moving so many units along the x-axis followed by so many units parallel to the y-axis and finally, so many units parallel to the z-axis. This operation does exactly the same job; it takes us from the foot of the angular momentum vector to its tip but in three steps, each of which involves a displacement along each of the three axes. Each of these three displacements is called a *component* of the angular momentum; in this case they are the x, y and z components. Any vector quantity, i.e. a quantity which is specified by a direction in space as well as a value or magnitude, can be resolved into three components, once a coordinate frame has been specified.

Enter the Magnetic Field

A magnetic field is another vector quantity; it has direction as well as magnitude; the direction is that in which a north magnetic pole would move and the magnitude is the amount of force which the pole experiences. Real magnetic fields will vary in strength

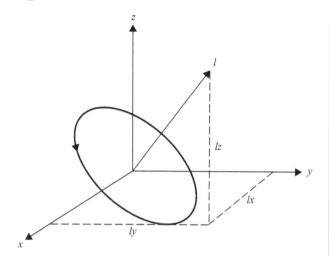

Figure 11.4. In an x, y, z coordinate frame, an electron's angular momentum vector l can be resolved or separated into three components l_x, l_y and l_z which are directed along the x, y and z axes.

and direction from one point in space to another but here it's vital to keep things simple; so let's assume that we have a magnetic field whose direction and strength are the same everywhere. The strength will be measured in gauss and we can conveniently arrange our coordinate frame so that the magnetic field points along the z-axis.

Now let's place in the magnetic field some atoms which are going to undergo electron transitions which here we'll assume will produce emission lines. An atom may contain a closed shell of electrons; the magnetic fields of these electrons effectively cancel each other out so they have no role to play here. To keep things simple let's assume that our atoms have just one optically active electron which may for example be in a level whose l value is 3. This l value determines the magnitude of the electron's angular momentum and indeed for our purpose here we can make the length of the angular momentum vector equal to the value of l itself, i.e. 3. Without the presence of the magnetic field this vector could point in any direction; the magnetic field though, interacts with the electron's magnetic field and this forces the electron's angular momentum vector (and of course its orbital plane) to orient itself in the field. However, only certain orientations are allowed under the rules of quantum mechanics and these orientations correspond to the values of m_l.

The orientation rule is a simple one; it says that the z component of the angular momentum vector (this of course is the angular momentum component which is parallel to the magnetic field) must equal one of the possible values of m_l; so in the case we're considering here this means -3, -2, -1, 0, 1, 2 or 3. Let's see how it works. The orientation can be along the direction of the field itself; i.e. along the z-axis and if this is the case, the electron is in the 'm_l equals l' sub-sublevel or $m_l = 3$. There are other possible orientations; one possibility is for the angular momentum vector to point towards the $-z$ direction (here $m_l = -3$) and yet another makes it point along the xy plane ($m_l = 0$); these orientations give us three of the seven m_l levels for $l = 3$. The remaining four orientations have angular momentum z components equal to -2, -1, 1 and 2 corresponding to the other allowed m_l values; in these cases the plane of the electron's orbit is tilted with respect to the direction of the magnetic field.

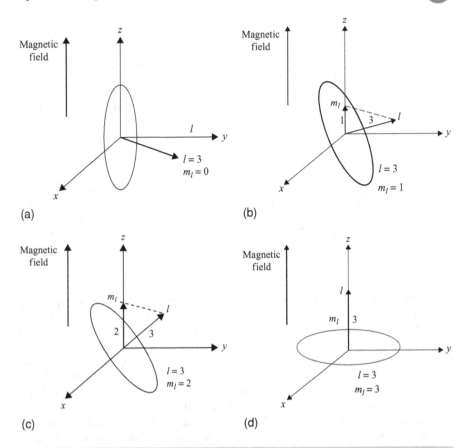

Figure 11.5. When an external magnetic field is present, the electron's orbit orients itself in the field so that the component of the angular momentum vector l which is directed along the field (here made to coincide with the z axis) is equal to one of the values of the magnetic quantum number m_l. A similar set of diagrams would show l pointing downwards here to correspond to negative values of m_l.

The $l = 3$ level is thus split into $7(2l + 1)$ m_l levels; each of these has a slightly different energy, however, the energy difference between two adjacent m_l levels is the same and this applies to all m_l levels for a given magnetic field. In fact the energy difference between two m_l levels depends directly on the strength of the magnetic field; for example double the magnetic field strength and the energy difference between the m_l levels also doubles.

There's another interesting feature here; quantum mechanics specifies how the angular momentum vector must orient itself with respect to the direction of the magnetic field. Here we have chosen this to be the z-axis; however that part of the angular momentum vector which lies within the xy plane can point in any direction. Indeed, the electron's angular momentum vector precesses around the direction of the magnetic field, just as the Earth's axis of rotation precesses around the pole of the ecliptic.

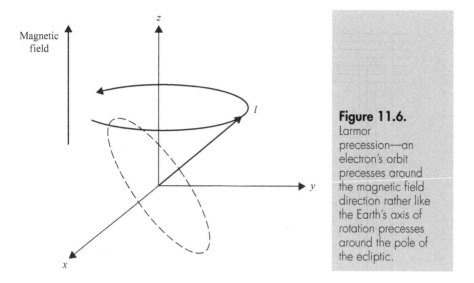

Figure 11.6. Larmor precession—an electron's orbit precesses around the magnetic field direction rather like the Earth's axis of rotation precesses around the pole of the ecliptic.

This phenomenon is called *Larmor precession* in honour of its discoverer the British physicist Joseph Larmor.

Electron Transitions in a Magnetic Field

Now let's take our electrons in the $l = 3$ level and see what happens when they 'jump' to the $l = 2$ level. In a transition the value of l changes from 3 to 2 so the l selection rule is obeyed. There is however a new quantum mechanical selection rule which says that in any transition m_l must also change by either $+1$, -1 or zero. If no magnetic field were present then the m_l levels would be indistinguishable; the electron would simply go from level $l = 3$ to $l = 2$ and the emitted photon would contribute to one single emission line in the final spectrum. The magnetic field together with the m_l selection rule means that there are now 15 possible transitions between the $l = 3$ and $l = 2$ levels; however because the m_l levels are equally spaced in energy, these 15 transitions group themselves in three groups of five where each transition within a group corresponds to the same change in m_l, the same change in energy and hence same wavelength of light emitted.

The three groups correspond to changes in m_l of -1, 0 and $+1$; in order of decreasing energy; the result is that the single spectral line is split into three by the magnetic field. This is called the *normal Zeeman effect*. The separation of the three components is determined by the energy separation of the m_l levels which in turn is determined by the strength of the magnetic field. The wavelength of the central component; i.e. the one which corresponds to no change in m_l has the same wavelength that the line would have in the absence of a magnetic field. There is however a very interesting twist to this tale.

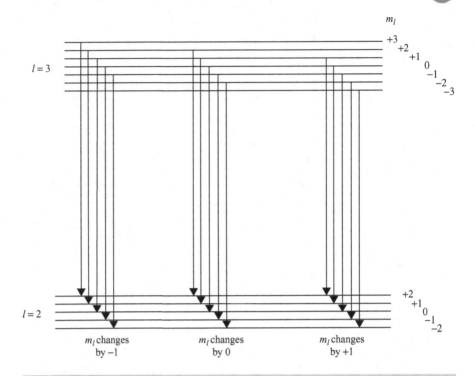

Figure 11.7. Here we see all possible transitions from the $l = 3$ to the $l = 2$ level. Each transition involves a change of m_l by either -1, 0 or $+1$. Because the m_l levels are equally spaced in energy, the transitions can be arranged into three groups which results in three spectral lines being observed. It's easy to show that this happens for transitions between any two l levels.

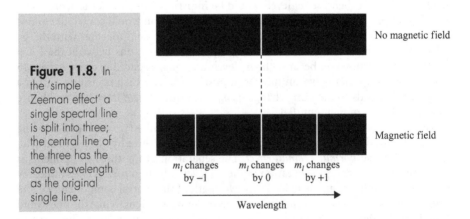

Figure 11.8. In the 'simple Zeeman effect' a single spectral line is split into three; the central line of the three has the same wavelength as the original single line.

Looking Straight Down the Magnetic Field

At the beginning of Chapter 2 we learned that accelerating electric charges (this includes oscillating charges) produce electromagnetic radiation. This is very much a classical physics idea but it can still be applied to quantum mechanical situations such as we have here. Before quantum theory was developed it was assumed that electric charges in atoms oscillated continuously thus enabling atoms to emit radiation continuously. The big difference with quantum theory is that there are no continuously oscillating charges in atoms; instead we have electrons which change their energy level giving rise to a brief, almost instantaneous emission of radiation called a photon. Even within quantum theory though, you can think of this energy change as resulting from the electron changing its state of motion around the nucleus and this amounts essentially to the electron suffering a very brief burst of acceleration.

Classical physics says that if an electric charge oscillates along a fixed line then the radiation comes out at right angles to this line; or more precisely the emitted wave has maximum amplitude in the direction at right angles to the line. As we move away from this direction the wave amplitude falls off and becomes zero in the direction along which the charge is oscillating. Quantum mechanics would say that in this situation the emitted photon is most likely to go in a direction at right angles to the direction of the acceleration and increasingly less likely to go in directions nearer to the acceleration direction. If we could perform an experiment to actually see what happens, we would indeed observe that the intensity of the emitted light falls off as we moved our position closer to the line of acceleration. Classical physics would interpret this as the result of many electric charges oscillating along the same line and all continuously producing exactly the same kind of electromagnetic wave. By contrast, quantum theory would interpret the results as due to large numbers of charges each emitting one single photon at a time, each of which has a different probability of coming out in different directions. The most important thing here is that if a charge, e.g. an electron, accelerates along a given line, the emitted light is most likely to come out at right angles to the line.

The magnitude of an electron's angular momentum vector is determined by the electron's orbital motion around the nucleus of the atom; if that motion changes in any way then the angular momentum will also change. The angular momentum vector's direction is always at right angles to the orbital plane but if we set up our x, y, z coordinate frame then we can think of the projection of the orbital plane onto the xy plane. The electron's motion around the nucleus is a combination of motion in the x, y and z directions so by projecting the orbital plane onto the xy plane we are thinking about the part of the electron's motion which is confined to this plane. The angular momentum for this part of the electron's motion is of course the z component, which is the part lined up with the magnetic field according to the rule described above.

Now think about an electron transition; as described above; m_l may change by -1, $+1$ or 0. A change in m_l means a change in the z component of the angular momentum and this can only happen if the electron's motion in the xy plane changes. A change of motion along the x direction will result in photons being emitted along both the y and

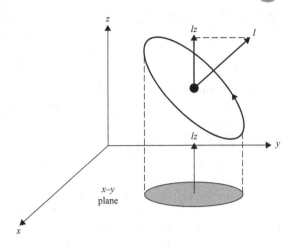

Figure 11.9. An electron's orbit can be projected onto the three planes of a coordinate frame as shown here for the xy plane. The z component of the electron's angular momentum results from the projection of the electron's motion onto the xy plane. The x and y components of the angular momentum are defined in a similar way.

z directions since these are both at right angles to the x direction. Similarly, a change of motion along the y direction will result in emission along the x and z directions. So if m_l changes as a result of a transition, photons can be emitted essentially in all directions. Consider now however the transition which results in no change in m_l; this means that the z component of the angular momentum remains constant. For

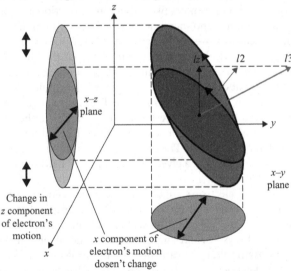

Figure 11.10. A transition takes an electron from the $l = 3$ level to the $l = 2$ level but with no change in m_l. This means that l_z remains constant; this in turn means there is no change in the x and y components of the electron's motion. I've shown this here for the 'x' component of the motion by projecting the electron orbits onto the xz plane. In a similar way, projecting the orbits onto the yz plane would show no change in the y component of the electron's motion. All of the change in the electron's motion which results from this kind of transition is thus directed along the z axis; i.e. along the direction of the magnetic field.

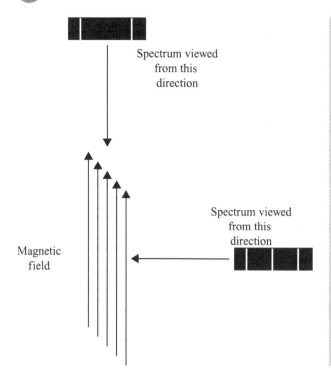

Figure 11.11.
When viewed in a direction at right angles to the direction of the magnetic field, the spectrum shows all three lines resulting from the Zeeman effect. When viewed along the direction of the field, the central line is missing.

Spectrum viewed from this direction

Spectrum viewed from this direction

Magnetic field

this to be so the electron's overall motion in the *xy* plane does not change, but the overall angular momentum has changed because the value of *l* has changed. This overall change in angular momentum can only be due to a change of motion or a brief acceleration along the *z* direction. The result is that photons are predominantly emitted at right angles to this direction, i.e. at right angles to the direction of the magnetic field; no photons are emitted along the direction of the field. The transition which involves no change in m_l is the one which produces the central line of the three; hence this central line is not observed when looking along the direction of the magnetic field. This is known as the *longitudinal Zeeman effect*.

Most text books describe this effect in terms of a pre-quantum mechanics model developed by the Dutch physicist H.A. Lorentz; this model gives the right answers but for the wrong reasons. It assumes that the emitted light is produced by continuously oscillating charges as described above. It does however also correctly predict the relative intensities of the various line components which are emitted in different directions. At right angles to the field the central component has twice the intensity of the other two; the 'missing bits' of these other two components are the portions which come out along the direction of the magnetic field.

How Wide Do the Lines Get Split?

Once again there is a simple 'plug the numbers in' formula for the normal Zeeman effect which we can use to give us some idea of how wide apart the components of a split line are. For wavelengths in angstroms and magnetic field strength **H** in gauss the formula gives the difference in wavelength $\Delta\lambda$ between the components which correspond to the m_l changes by $+1$ or -1 transitions and that which comes from the m_l changes by 0 transition.

$$\Delta\lambda = 4.67 \times 10^{-13} \times \lambda_0^2 \times H \qquad (11.1)$$

One thing to note here is that the separation of the components depends on the wavelength squared; this means that any effect will be much greater for longer wavelength lines.

Let's try this out on a couple of examples; as usual we'll use the Hα line at λ6563. First, take the example of a sunspot with a magnetic field of 3000 G; Eq. (11.1) gives us

$$\Delta\lambda = 4.67 \times 10^{-13} \times (6563)^2 \times 3000 = 0.06\,\text{Å}$$

This isn't much of a separation and high-resolution spectroscopy would be needed to show it. A more promising candidate might be the star AM Herculis; this is the prototype of a class of interacting binaries called polars, whose white dwarf components have very powerful magnetic fields in the region of 10 million gauss; such fields would produce a separation of 200 Å.

Complex Atoms

We kept things very simple above when describing the Zeeman effect; in particular we considered atoms with only one optically active electron and energy levels which in the absence of a magnetic field are simple so-called singlet levels. Towards the end of Chapter 3 we saw how atoms which contain a net surplus of electrons with the same spin produce much more complicated spectra in which the energy levels are split into triplets, quartets, etc. The magnetic fields resulting from the electrons' spins interact as do those from the orbital motions to modify and separate the levels. This means that for atoms like these there are many more levels to be further split by the action of a magnetic field; this results in many more possible transitions and results in a single spectral line being split into possibly many more than three components. This effect was called the *anomalous Zeeman effect* because Lorentz's classical model could not explain it. With the discovery of electron spin, it was seen to be a very natural effect, though the name has stuck.

Very Strong Magnetic Fields

The almost delicate interplay between the magnetic fields of several electrons in complex atoms gets completely disrupted (mercifully some would say) by a very powerful magnetic field. The complicated separation of the lines reverts back to the simple three components of the normal Zeeman effect; this itself constitutes yet another effect called the *Paschen Bach effect*.

Summary

- Orbital motion produces angular momentum.
- Angular momentum is a vector quantity; it has direction as well as magnitude.
- For orbital motion the angular momentum vector points at right angles from the centre of the orbit.
- If the orbital motion is seen to be anticlockwise, the angular momentum vector points towards the observer.
- In a magnetic field an electron's angular momentum and orbital plane are oriented at fixed angles to the field which correspond to the magnetic quantum number m_l.
- In a transition m_l must change by -1, $+1$ or 0.
- The possible changes in m_l split a spectral line into three components; this is called the normal Zeeman effect.
- The central line components is seen best at right angles to the magnetic field direction and not at all along the direction of the field. This is known as the longitudinal Zeeman effect.
- Complex atoms can give rise to the anomalous Zeeman effect in which a line splits into more than three components.
- Very strong magnetic fields disrupt the magnetic field structure for complex atoms, resulting in a reversion to the normal Zeeman effect. This is called the Paschen Bach effect.

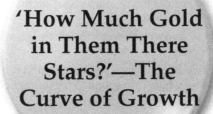

'How Much Gold in Them There Stars?'—The Curve of Growth

At the beginning of this book I recounted August Comte's classic claim that chemical knowledge of the stars was something that mankind would never possess. Unfortunately for Monsieur Comte, not only have astronomers learned to recognise different chemical elements in stars, they have also often managed to determine *how much* of each element is present. The actual amount of a given element which is present in the atmosphere of a star or indeed throughout the Universe as a whole is called its *abundance*, so let's first see how the abundance of an element is quantified.

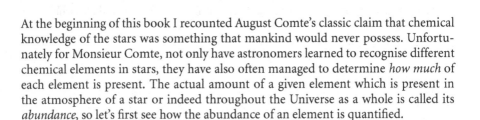

Abundances

One often reads in popular astronomy books these days that about 75% of the Universe is hydrogen; about 25% is helium and less than 1% is everything else. These figures are basically true of course but only if we're talking in term of mass, i.e. so many tons of hydrogen to so many tons of helium, sodium and so on. What's often of more interest to the spectroscopist though is the actual number of atoms which are present because atoms in a stellar atmosphere knock out photons and more atoms mean that more photons get knocked out which results in a deeper darker absorption line. If we talk about abundances in terms of number of atoms, then the Universe is more like 91.5% hydrogen to 8.5% helium and even less of everything else. Both of these ways of quoting abundance are equally valid however provided you state which one you're using.

Whichever way we choose to talk about abundances, hydrogen is by far the most common element and so the abundance of every other element is almost always referred to that of hydrogen; so element abundances are in fact *relative abundances*;

that is, relative to hydrogen. Abundances which are given in terms of number of atoms will always be less than those given in terms of mass. An atom of any element will always be more massive than the mass of a hydrogen atom; so as we move through the periodic table it takes fewer atoms to equal a given fraction of a standard mass of hydrogen. Abundances in both systems are given as what at first seem like rather small numbers; the highest one being 12.00 for hydrogen itself. Typical values for helium are 10.93 if we're dealing with numbers of atoms and 11.53 if we're talking about mass; let's see how this system works.

Take first the figure of 11.53 for the mass abundance of helium and subtract this from 12.00 (the hydrogen figure) to get 0.47. If you now fancy a trip down memory lane, dig out that old book of log tables which you used in high school and look up the antilog of 0.47. Alternatively you could use your calculator; use the 'invert' and 'log' keys (log here is log to base 10 *not* natural log or log to base 'e') followed by 0.47 or for yet another method enter the number 10 followed by the 'x^y' key followed by 0.47. Each of these will give you the answer as 2.95. This tells us that in terms of mass there are 2.95 times as much hydrogen in the Universe as there is helium. This clearly fits in with the well known 'percentage abundances' of about 75 to 25%.

Finally, let's see if we can answer the question posed in the title of the chapter; we'll do this just for the number of atoms involved. The abundance of gold in terms of number of atoms is 0.6; so again subtracting from 12.00 gives us 11.4 this time. Taking the antilog of this (book of tables or calculator) will give the answer as 2.5×10^{11}; this is the number of times that hydrogen atoms outnumber those of gold, so for every 250 billion atoms of hydrogen in the Universe, there's just one of gold. Some stars may contain slightly more gold than this but it's pretty rare stuff; if you find it in your backyard, then truly your backyard is a very exotic part of the Universe.

A Laboratory Experiment

Back in Chapter 5 we learned that the best way to quantify the intensity of a spectral line is in terms of its equivalent width; i.e. the width in angstroms of an artificial saturated line having a rectangular profile of area equal to that of the real line profile. What we want to find out is how the equivalent width of a spectral line changes, as we increase the number of atoms which are producing it. To do this let's imagine doing a laboratory experiment with hydrogen to see how the equivalent width of our old friend the Hα line changes, as we increase the amount of hydrogen which is present.

We'd need a source of continuum radiation (basically a very bright light bulb) to simulate the photosphere of a star; we'd also need some kind of enclosure to contain the hydrogen and which could be heated to the kind of temperature needed to excite the hydrogen atoms to the $n = 2$ level, i.e. about 10,000 K. We'd also need some mechanism for increasing and measuring the density of the hydrogen in the enclosure and finally a spectroscope to observe the spectrum and measure the equivalent width.

At relatively low densities, the line profile would be the basic Voigt profile described in Chapter 5, i.e. a combination of a naturally broadened profile and a thermal Doppler broadened profile. As we increase the density of hydrogen, the line profile

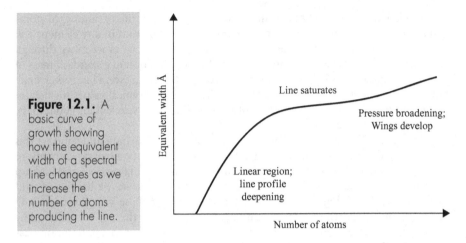

Figure 12.1. A basic curve of growth showing how the equivalent width of a spectral line changes as we increase the number of atoms producing the line.

would deepen as increasing numbers of atoms knock out more of the incoming photons with wavelengths in the vicinity of λ6563. The line would also broaden slightly as the number of atoms with the highest radial velocities increased. Eventually the line core would reach the bottom of the continuum and the line would become saturated. After this stage, aside from a very slight increase in core width, increasing the number of hydrogen atoms would produce virtually no effect on the line profile because having become saturated, the profile has 'nowhere to go' as it were. With still further increase in density though the effects of pressure or collision broadening begin to set in; the line profile wings develop and these enable the equivalent width to increase once more.

If we now plot our results in the form of a graph of equivalent width against number of atoms, we see that for low densities the equivalent width increases in direct proportion to the number of atoms which is pretty much what we'd expect. The graph then almost levels off as we reach the line saturation zone and eventually turns upwards again as collision broadening sets in. The graph that we have produced is called a *curve of growth*; in this case for the Hα line. The part of the graph which we are interested in here is the so-called 'linear region', where the equivalent width increases in direct proportion to the number of atoms; this is shown in Fig. 12.1.

A Bit of Theory

Imagine the usual vast population of hydrogen atoms with their electrons in the $n = 2$ level; this population sits in a radiation field which is capable of producing a spectrum showing all the optical Balmer lines. The Balmer lines themselves of course result from transitions from the $n = 2$ level to the $n = 3, 4, 5$, etc. levels. Because our radiation field is capable of producing all of these transitions, it would be natural to presume that the equivalent widths of all the resulting Balmer lines would be the same. However this is not so; we would observe that the equivalent width of the Hα line was the largest followed by that for the Hβ line and so on.

Take a hydrogen atom with its electron in the $n = 2$ level; an incoming $\lambda 6563$ photon can excite the atom to the $n = 3$ level. *This however isn't guaranteed*; as with all things to do with quantum mechanics there's a finite probability that the photon will be absorbed, but also a chance that it will be 'ignored' by the atom. An incoming $\lambda 4861$ photon (the wavelength of the Hβ line) has an even lower probability of doing its job; namely exciting the atom to the $n = 4$ level and this trend of decreasing probability carries on through the Balmer series. So for a given number of atoms the Balmer lines get progressively weaker from Hα through Hβ and so on.

There are three ways in which this chance or probability of a photon being absorbed, is expressed; they are all directly connected to each other but they all occur in the literature. The first way is not surprisingly called the *transition probability* and for absorption this is denoted by the letter B with two subscripts which denote the lower and upper levels of the transition; so the transition probability for the Hα line would be written B_{23}; that for the Hβ line, B_{24} and so on. There are also transition probabilities for downward, i.e. emission transitions which are written A_{32}; A_{42} etc.; and finally there are transition probabilities for what's known as *stimulated emission*. This phenomenon can happen when an incoming photon can actually cause an electron to drop down from a higher level to a lower level; stimulated emission is the process that drives lasers and masers. These transition probabilities are different for different atoms and for different transitions within the same atom; they are also called *Einstein coefficients* because it was he who first conceived them (one of the great man's non relativity contributions to twentieth-century physics). Their values which are higher for more probable transitions are either determined by doing spectroscopy lab experiments or by very clever people who enjoy doing quantum mechanical calculations.

Another term directly related to the Einstein coefficients is the *line strength* and it's important to mention it here, because if you came across it in the literature you could be forgiven for thinking that this term means the same thing as the equivalent width. As with the Einstein coefficients, line strengths are just another way of expressing the chance or likelihood of a transition occurring. The third way of quantifying the probability of a transition gets its name from classical or pre-quantum physics; this is the *oscillator strength* and it's used quite a lot by astronomers especially in relation to the curve of growth. Before quantum mechanics it was believed that atoms absorbed radiation as a result of electric charges within the atom (i.e. electrons) being made to oscillate; the oscillator strength is simply a quantum mechanical way of expressing the likelihood that this will happen. This is essentially entirely equivalent to the more correct interpretation given by the Einstein coefficients but we'll stick with it here, because as mentioned above, astronomers tend to use the oscillator strength when dealing with the curve of growth. The oscillator strength is always denoted by the letter 'f' and again if we're specifically referring to a particular transition we use subscripts; so we write f_{23} for the Hα oscillator strength and so on.

A Second Laboratory Experiment

Let's do our hydrogen experiment again but this time we'll measure the equivalent widths of all the Balmer lines; not just the Hα line. We'll also keep the density of hydrogen in the enclosure fixed and low enough that the lines are not saturated. What

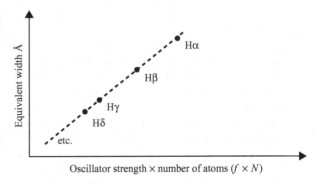

Figure 12.2. A curve of growth (the linear region) for the Balmer series of hydrogen. Each line in the series produces a different point in the plot because it is produced by a different number of atoms.

we can't do now of course is plot equivalent width against number of atoms for a single line, but if we know the oscillator strengths for the Balmer line transitions, what we can say is that for our fixed number of N hydrogen atoms, $f_{23} \times N$ of them will make the Hα line; $f_{24} \times N$ will make the Hβ line and so on. Now instead of having the equivalent width of just one line for a range of number of atoms, we have instead the equivalent widths of a series of lines, each for a different number of atoms; the number of atoms in each case being given by the oscillator strength for the transition multiplied by the total number of atoms. This time we plot equivalent width against $f \times N$; N is fixed of course bur f varies as we move through the members of the line series and so $f \times N$—the number of atoms producing each transition, varies. Once again we have a curve of growth for hydrogen but this time we are using all the members of the Balmer series to produce it. The reason for doing things this way is that we can't change the number of atoms of any element including hydrogen in the outer layers of a star, but we can observe and determine the equivalent widths of whole series of lines and plot curves of growth, provided we know the relevant oscillator strengths; as mentioned above these have been determined for many transitions by calculation based on quantum mechanics or by laboratory experiments.

Another Bit of Theory

Multiplying the total number of atoms by the oscillator strength for a transition gives the number of atoms which produce the spectral line corresponding to the transition and this of course determines the equivalent width of the line. The 'total number of atoms' here though means the number which are in the right energy level for producing the transition series we are observing. For the Balmer series, this means those with electrons in the $n = 2$ level of course. However, in our vast population of atoms there will be many whose electrons are in other energy levels, together with possibly a large number of ionised atoms if the temperature is high enough. If we now take the 'total number of atoms' to mean *all* atoms of a given element, then we need to multiply this by a number which gives us first the fraction of atoms which are not ionised and then by a number which gives us the fraction of these neutral atoms which are in the required energy level. Finally, as above, we multiply this number of atoms by the oscillator strength for the relevant transition. This final number will determine the equivalent width of the line which gets plotted on the curve of growth.

Astronomers use two equations whose mathematical details we won't go into here; the first is called *Saha's equation*. This gives the fraction of atoms which are either ionised or neutral; in addition, for more complex atoms it can give the fractions of atoms which are in various states of ionisation. All of these fractions depend of course on the temperature of the gas, but they also depend on the ionisation potential for atoms with one optically active electron; or the ionisation potentials for succeeding levels of ionisation in multi-electron atoms. Having calculated the number of neutral atoms (or indeed the number of atoms for an element in a given state of ionisation, if we are interested in spectral lines due to ionised atoms) we now need to calculate the fraction of these which are in a given energy level. This is done using *Boltzmann's equation*; again the number of atoms in a given level depends on temperature, but this time also on the excitation potential for the required energy level, i.e. the energy in electron volts needed to raise the electron to that level.

Summarising; we start with the total number of atoms N and multiply by a term derived from Saha's equation (let's call this 'S') to give us the number of neutral atoms; if the temperature is sufficiently low, then all the atoms will be neutral and S will simply equal 1. We then multiply the number of neutral atoms by the term derived from Boltzmann's equation which we can call 'B', to give the number of atoms in the appropriate energy state and finally we multiply this by the oscillator strength f for each transition to give the number of atoms which contribute to each spectral line in a series. The equivalent width which we can call 'W' will be proportional to this number, provided the lines are relatively weak, i.e. not saturated; so we can say

$$W = A \text{ constant} \times S \times B \times f \times N \tag{12.1}$$

A (Very Small) Bit of Mathematics

Here are two very simple facts about logarithms:

- The logarithm of the product of two or more numbers is equal to the sum of the logarithms of the individual numbers. So

$$\log(A \times B) = \log(A) + \log(B) \tag{12.2}$$

- The logarithm of the ratio of two numbers is equal to the difference of the logarithms of the individual numbers. So

$$\log(A/B) = \log(A) - \log(B) \tag{12.3}$$

If we convert Eq. (12.1) into logarithms we can write it like this

$$\log(W) = \log(\text{constant} \times S \times B \times f) + \log(N) \tag{12.4}$$

Armed with this vital bit of mathematics, we can now start to determine element abundances.

Determining Abundances

It probably goes without saying, that in order to determine the relative abundance of some element in the outer layers of a star we first need to be able to identify the spectral lines for that element and also the transition series to which they belong. This will enable us to look up the relevant ionisation and excitation potentials together with all the relevant oscillator strengths. We would also do the same thing for the Balmer lines in order to determine the relative abundance for the element. As indicated above, we'll also confine our attention to relatively weak unsaturated lines so that we can use Eq. (12.4).

The first thing to do is to determine the equivalent widths of the Balmer lines and the element lines and convert these into logarithms. This gives us two sets of values for $\log(W)$. Now have a look at Eq. (12.4); the 'constant' on the right-hand side has only a trivial scaling effect on the curve of growth plot and we can ignore it. The factors S and B which come from the Saha and Boltzmann equations depend in part on the temperature, which we may not have reliable information about; however we can assume that within the part of the star's atmosphere where the lines are formed, it is constant; so again we can ignore it. Ionisation potentials, excitation potentials and oscillator strengths can be looked up from standard reference sources. The remaining big unknown of course is 'N' the total number of atoms; but this is the number we're trying to find and oddly, just for the moment we ignore it.

We now produce two curves of growth by plotting $\log(W)$ against $\log(S \times B \times f)$; one for the hydrogen lines and one for the element lines. The result will be two sloping straight lines which are parallel to each other but displaced relative to each other along the horizontal axis. A line drawn parallel to the horizontal axis which crosses both plots corresponds to equal equivalent widths; if we denote 'H' to stand for hydrogen and 'E' to stand for the other element then Eq. (12.4) tells us that on this line

$$\log(S \times B \times f)_{\mathrm{H}} + \log(N)_{\mathrm{H}} = \log(S \times B \times f)_{\mathrm{E}} + \log(N)_{\mathrm{E}} \qquad (12.5)$$

Figure 12.3.
Plotting curves of growth for hydrogen and some given element in this way exploits Eq. (12.6); the horizontal displacement of the two plots gives the relative abundance of the element.

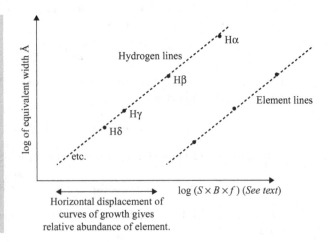

A bit of swapping round gives us

$$\log(S \times B \times f)_H - \log(S \times B \times f)_E = \log(N)_E - \log(N)_H \qquad (12.6)$$

The bit on the left of the equal sign is just the horizontal displacement of the two curves of growth and by using the vital Eq. (12.3), the right-hand side becomes

$$\log(N)_E - \log(N)_H = \log(N_E / N_H) \qquad (12.7)$$

The left-hand side here will always be negative because hydrogen is more abundant than any other element; this means that the left-hand side of Eq. (12.6) will also be negative. This results from the fact that the curve of growth for any element will always be to the right of that for hydrogen in a curve of growth plot, because the equivalent widths of the element lines will be less than those for the hydrogen lines.

As an example, suppose the horizontal displacement of our two plots equals -4; by convention $\log(N)_H$ always equals 12 so in this case $\log(N)_E$ must equal 8. The abundance of our element 'E' would then be given as 8.

This then in principle is how it's done; overall cosmic abundances have been pretty well worked out, although abundances among different stars and for that matter in objects like planetary nebulae can vary. The result is that abundance determination using curve of growth methods is still a fairly active area of research.

Summary

- Element abundances are usually listed as logarithms relative to hydrogen which has a value of 12.00.
- The increase of optical depth of a spectral line with the number of atoms producing it is called a curve of growth.
- Using stellar spectra, a curve of growth for a spectral series from some element is plotted alongside one for hydrogen; the horizontal displacement of the two plots gives the relative abundance of the element.

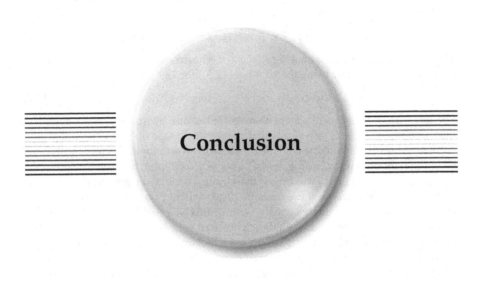

Conclusion

This has been a theory book for observational amateur astronomers. This is perhaps a bit unusual because most astronomy 'theory books' tend to be written for armchair astronomers and they tend to deal with fairly exotic subjects like black holes and cosmology. Spectroscopy however is different; for one thing it's a fairly new area of research for amateurs, so until now there hasn't been much of a compelling need for a book like this. The other thing is that spectroscopy by its very nature involves a lot of physics and most amateur astronomers don't have physics degrees. There was, I reckon, an urgent need to provide an explanation of some of the theory behind spectroscopy, including the relevant physics. That was the main idea behind this book.

I hope that by now you've gained a deeper insight into spectroscopy and most of all come to realise (if you didn't realise already) that spectroscopy involves far more than just identifying lines in spectra. Indeed, identifying spectral lines is very likely the subject for another book; one which needs to be written by someone with many years of observational experience. Maybe out there, there's a kind hearted professional astronomer or a pioneering amateur who could do this. If, as a result of reading this book however, you can better appreciate the things which are going on in your spectra, then I reckon I've done my job.

Clear skies and good luck with your spectra.

Appendix A— Powers of Ten

First take as an example the number 2.9 and multiply it by 10: we get 29 of course

$$2.9 \times 10 = 29$$

Similarly

$$2.9 \times 100 = 290$$
$$2.9 \times 1000 = 2900 \text{ and so on.}$$

We are simply moving the decimal point one place to the right each time. Now take

$$2.9 \times 100,000,000,000$$

Here we have to make a conscious effort to count the zeros and for each one of them, move the decimal point one place to the right. So we get

$$290,000,000,000$$

This is not a very elegant way to write two hundred and ninety thousand million. A much better way is to use *scientific notation*.

$$100 = 10 \times 10, \text{ i.e. 10 multiplied by itself twice}$$
$$1000 = 10 \times 10 \times 10, \text{ i.e. 10 multiplied by itself 3 times}$$
$$10,000 = 10 \times 10 \times 10 \times 10, \text{ i.e. 10 multiplied by itself 4 times and so on.}$$

Scientific notation writes these numbers like this

$$100 = 10^2$$
$$1000 = 10^3$$
$$10000 = 10^4 \text{and so on.}$$

You can see the advantage of writing a number like 100,000,000,000, i.e. one hundred thousand million as 10^{11}. You pronounce this 'ten to the eleven' A number like 10^{11} is also referred to as '10 to the power 11' or '10 raised to the power 11.' Back to our number 2.9; now instead of writing $2.9 \times 100,000,000,000$, we can simply write this as

$$2.9 \times 10^{11}$$

This is an example of scientific notation. The 10 raised to the power bit is called the *exponent*.

Let's take our 2.9×10^{11}

It comes in two parts: 2.9 and 10^{11}

We could multiply the 2.9 by 10 to get 29 and then we would have

$$10 \times 2.9 \times 10^{11} = 29 \times 10^{11}$$

However, you would never write this number like this. The first part of a scientific notation number should always be greater than 1 and less than 10. We keep the first bit as 2.9 and so we multiply the 10^{11} bit by 10 instead.

$10^{11} \times 10$ becomes 10 multiplied by itself 12 times, i.e. 10^{12}. So our scientific notation number is written as

$$2.9 \times 10^{12}$$

The '11' in 10^{11} is often referred to as the index of the power of 10. Notice above that when we multiplied 10^{11} by 10 we simply increased the value of the index by 1 to get 10^{12}. Had we multiplied 10^{11} by 100 (i.e. 10^2) we would have got 10^{13}. Multiplying by 1000 (10^3) would have given us 10^{14} and so on. In other words, when we multiply two powers of 10 together, we simply add the two indices; e.g.

$$10^7 \times 10^{11} = 10^{18} \text{etc.}$$

Just as 10^2 is the number 1 with 2 zeros after it and 10^3 is the number 1 with 3 zeros after it etc. the number 10 on its own is the number 1 with 1 zero after it, i.e.

$$10 = 10^1;$$

and the number 1 is of course 1 with no zeros after it, i.e.

$$1 = 10^0.$$

Now let's multiply together, two numbers in scientific notation, so let's work out

$$2.9 \times 10^7 \times 3.6 \times 10^5$$

First, multiply the powers of 10

$$10^7 \times 10^5 = 10^{12}$$

Using our calculator

$$2.9 \times 3.6 = 10.44$$

So this would initially give us 10.44×10^{12} but remember we keep the first part of the number less than 10. So we divide it by 10 to give us 1.044 and we must multiply the exponent (the powers of 10 bit) by 10 to balance the books. So we get

$$1.044 \times 10^{13}$$

Now let's try dividing two powers of 10; let's try

$$10^5/10^2$$

This of course is just 100,000/100—the zeros cancel out to give 1000. 1000 is of course 10^3, but notice that $3 = 5 - 2$. In other words, when we divide one power of 10 by another, we simply subtract the lower power from the upper power, e.g.

$$10^{12}/10^7 = 10^5$$

So far we've talked about a power of 10 in terms of the number of zeros after the '1' by which we mean of course the number of zeros to the right of the '1' and we get a positive index for our power of 10. So if we think of a zero to the right of the '1' as contributing to a positive index for the power of 10, we could think of a zero to the left of the '1' as contributing to a negative power of 10 index. Thus the number 0.1 (written this way as opposed to .1) has one zero to the left of the '1' and in fact

$$0.1 = 1/10 = 10^{-1}$$
$$0.01 = 1/100 = 1/10^2 = 10^{-2} : \text{the '1' has 2 zeros to its left}$$
$$0.001 = 1/1000 = 1/10^3 = 10^{-3} : \text{the '1' has 3 zeros to its left}$$

So for example $0.000001 = 1/1000000 = 1/10^6 = 10^{-6}$

This is consistent with the rules for multiplying and dividing powers of 10, e.g.

$$10^7 \times 10^{-4} = 10^7 \times 1/10^4 = 10^7/10^4 = 10^{(7-4)} = 10^3$$

Summarising so far, we can use scientific notation with a positive power of ten index to very conveniently represent large numbers; e.g. 2.9×10^{18} and by using a negative power of ten index, we can neatly represent very small numbers, e.g. 2.9×10^{-18}. What about square roots and cube roots and so on? Can we use scientific notation here?

100 is 10×10 which is 10^2 of course. 10 is the square root of 100. But 10 is of course 10^1. So to get the square root of 100 we divided the index on 10^2 by 2; i.e $\sqrt{10^2}$ is simply $10^{2/2}$ which equals 10^1, i.e. 10.

The square root of 10^4 is 100, i.e. 10^2. Once again to get the square root, we divide the power of ten index by 2.

10 is the same as 10^1 and so by the same token, to get the square root, we divide the power of ten index by 2. So $\sqrt{10} = \sqrt{10^1} = 10^{1/2}$

So $\sqrt{10} = 10^{1/2}$

In a similar way the cube root of 10 is given by $10^{1/3}$; the fourth root by $10^{1/4}$ etc.

Now take for example the number $(10^4)^6$; this is *not the same* as $10^4 \times 10^6$, it is the number 10^4 multiplied by itself 6 times. Do this longhand and you can see that you get a 1 with 24 zeros after it; i.e.

$$(10^4)^6 = 10^{(4 \times 6)} = 10^{24}$$

Now note that a number like $(2.9 \times 10^4)^6$ is equal to $(2.9)^6 \times 10^{24}$

Right! Let's do a full-blown formula in scientific notation. Back in the 1970s, theoretical physicists were thinking about an interval of space which was so small that Einstein's general theory of relativity breaks down and is replaced by a still unknown quantum theory of gravity. How big would this interval be? The formula they arrived at is given by

$$L = \left[\frac{Gh}{c^3}\right]^{1/2}$$

L is called the Planck length, G is the universal constant of gravitation $= 6.67 \times 10^{-11}$, h is Planck's constant $= 6.6 \times 10^{-34}$, c is the speed of light $= 3 \times 10^8$
So we have

$$L = \left[\frac{6.67 \times 10^{-11} \times 6.6 \times 10^{-34}}{(3 \times 10^8)^3}\right]^{1/2}$$

Let's sort out the powers of 10 first

$$10^{-11} \times 10^{-34} \text{ gives us } 10^{-45}$$
$$(10^8)^3 = 10^{24}$$
$$10^{-45}/10^{24} = 10^{-69}$$

Now the number bits

$$6.67 \times 6.6/3^3 = 44.022/9 = 4.89$$

So far we have 4.89×10^{-69} but we need the square root of this. To get the square root of 10^{-69} we need to divide the index of the exponent by 2 but this is awkward unless we have an even numbered index. So, breaking the rule above (we are after all in the middle of a calculation and we'll be sure to write the number correctly at the end) let's write 4.89×10^{-69} as 40.89×10^{-70}, i.e. we've multiplied the number bit by 10 and divided the power of 10 by 10.

The square root of 10^{-70} is 10^{-35} and the square root of 40.89 is (use calculator) 6.39.

So our final answer in correct scientific notation is

$L = 6.39 \times 10^{-35}$ and this distance is in metres by the way—a very tiny distance indeed.

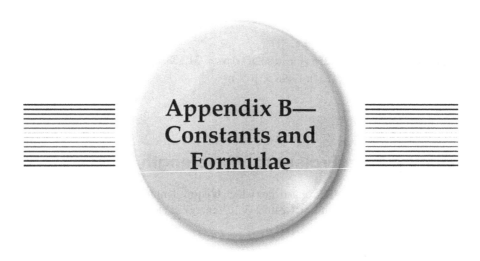

Appendix B—Constants and Formulae

I've gathered together here some of the more important physical constants and useful 'plug the numbers in' formulae which I've used in the book. As well as enabling you to easily calculate useful numbers, these formulae are the key to exploring some piece of spectroscopic theory and getting a real feel for what's going on.

Physical Constants

These are in what's known as the MKS or metre (m) kilogram (kg) second (s) system; also known as the SI system.

Speed of light c: 2.998×10^8 m/s.
Planck's constant h: 6.626×10^{-34} J s.
Gravitational constant G: 6.673×10^{-11} N m^2/kg^2
Boltzmann constant k: 1.381×10^{-23} J/K
Mass of hydrogen M_H: 1.674×10^{-27} kg
One electronvolt 1 eV: 1.602×10^{-19} J
One angstrom 1 Å: 10^{-10} m

Astronomical Constants

One Solar mass: 1.99×10^{30} kg
One (sidereal) year: 365.256 (mean solar) days $= 31,558,118$ s.
One astronomical unit (AU): 1.496×10^{11} m

Formulae

Energy E Equivalent to Wavelength λ

This gives the wavelength in angstroms which is equivalent to energy in electronvolts.
  Wavelength in angstroms $= 1.24033 \times 10^4$/energy in electronvolt (see Chapter 3).

Doppler Formula

Wavelength change $\Delta\lambda$ in terms of velocity v

$$\Delta\lambda = \lambda \times v/c$$

  Velocity in terms of wavelength change

$$v = c \times \Delta\lambda/\lambda$$

$\Delta\lambda$ and λ are in angstroms; v must be in the same units as c; e.g. metres per second.
or kilometer per second.

Relativistic Doppler Formula

$$\Delta\lambda = \lambda \times \frac{1 + \dfrac{v}{c}}{\sqrt{1 + \dfrac{v^2}{c^2}}} - 1$$

There's probably little need to use this for velocities less than about 5000 km/s unless
you're doing high-resolution spectroscopy (see Chapter 4).

Kepler's Third Law Formula

Orbital period P (e.g. of a binary) in terms of binary separation a and stars' masses
M_1 and M_2;

$$P^2 = \frac{a^3}{M_1 + M_2}$$

P is in (Earth) years, a in astronomical units (AU) and M_1 and M_2 are in solar masses
(see Chapter 4).

Full Width Half-Maximum (FWHM) Formula

The FWHM is the total width in angstroms of a Doppler (or thermally) broadened spectral line at half its maximum intensity (for an emission line) or depth (for an absorption line). The temperature of the gas in Kelvin is given by

$$T = 1.968 \times 10^{12} \times (\text{FWHM})^2 / \lambda_0^2$$

λ_0 is the rest wavelength in angstroms of the spectral line (see Chapter 5).

Wien's Displacement Law

This gives the wavelength $\lambda_{\max}$ in angstroms at maximum emission for a black body (a star is a reasonable approximation to this) of temperature T Kelvin.

$$\lambda_{\max} = 28,978,200/T$$

See Chapter 8

Zeeman Effect Formula

This gives the separation $\Delta\lambda$ in Angstroms between each of the three components of a spectral line of wavelength λ_0 which is split by a magnetic field of strength H gauss (see Chapter 11)

$$\Delta\lambda = 4.67 \times 10^{-13} \times \lambda_0^2 \times \mathbf{H}$$

Index